# Molecular Characterization of CA125 in Benign and Malignant Ovarian Tumors Using Immunoradiomatric *Assay Technique*

**Prof. Dr. Sami A.AL-Mudhaffar**

**Dr. Majid Karbon**

# Contents

# CHAPTER TWO
## Preliminary Binding Studies of CA125 to $^{125}$I-anti CA125 antibody in Ovarian Tumor Homogenates.

# CHAPTER THREE

# Chromatography Purification of CA125 by Gel Filtration and Binding Characterization to its Specific Antibody.

# CHPTER FOUR
## Kinetic & thermodynamic studies of the binding of CA125 with $^{125}$I-antiCA125 antibody in ovarian tumor homogenates 96

# CHPTER FIVE

## Spectroscopic Studies on Isolated ($^{125}$I-anti CA125 antibody/CA125) Complex

## Summary

Immunoradiomatric assay technique was used to measure the level of serum tumor markers CA125 and CEA in 30 healthy women, 24 patients with benign ovarian tumors and 34 patients with malignant ovarian tumors. Upon comparison of the results, the CA125 level was found above normal ($35u.ml^{-1}$) in 80% of patients with ovarian cancer, while CEA level was elevated in 44% of patients with ovarian cancer (more than $3.0 \ u.ml^{-1}$). Among the pre-menopausal women, the values of CA125 and CEA in sera of patients with malignant ovarian tumors were found to be significantly higher ($P<0.05$) than their corresponding values in sera of healthy and that in patients with benign tumors. The mean value of CA125 tumor marker in relation to menopausal status of ovarian cancer women, was found to be significantly higher ($P<0.05$) in post-menopausal patients, than that in pre-menopausal patients, while there were no differences found in the CEA value between the two statuses.

Modified immunoradiometric assay (IRMA) was used to characterize the binding of $^{125}$I-antiCA125 antibody to CA125 in the supernatant fraction of ovarian tumor homogenates. Different factors affecting this binding were extensively studied such as pH, time, temperature, concentration of salts, concentration of antibodies and concentration of antigens

CA125 of homogenate obtained from post-menopausal patients with malignant ovarian tumor was partially purified by gel filtration technique. Two forms of CA125 were found (BI and BII) with molecular weights 670 and 100 kDa respectively. The binding characteristics of the partial purified CA125 with $^{125}$I-antiCA125 antibody were investigated.

Kinetic parameters of the binding of $^{125}$I-antiCA125 antibody with CA125 of homogenate obtained from pre-menopausal patients with benign and malignant tumors, post-menopausal patients with malignant tumors, and partially purified CA125, at different temperatures were determined. The results indicated that the binding reaction was time and temperature dependent process and it follows second order rate law in all studied groups.

The thermodynamics of the binding of $^{125}$I-antiCA125 antibody with CA125 of homogenate obtained from pre-menopausal patients with benign and malignant tumors, post-menopausal patients with malignant tumors, and the partially purified CA125, at different temperatures were determined. All studied groups showed exothermic and spontaneous reactions and the participation of enthalpy was low. Thermodynamic parameters of transition state ($\Delta H^*$, $\Delta G^*$, $\Delta S^*$) and activation energy (Ea) were also determined.

Spectroscopic studies, in the U.V region, were carried out on the complex ($^{125}$I-antiCA125 antibody/CA125) formed of the partially purified CA125, from post-menopausal patients with malignant ovarian tumors, to its specific antibody. Different factors affecting the absorption band such as pH, polarity and denaturation agents were extensively studied. The heat stability and spectrophotometric pH titration of the complex were also included in this work.

# Chapter One

# Introduction
# &
# Literature Survey

# Introduction and Literature Survey:

## 1.1 Historical background of tumor marker

Bence jones protein which is exhibits unusual solubility in water was the first tumor marker identified in 1848, it appears in large amounts in the urine of patients with multiple myeloma. More than 100 years after its discovery the Bence-Jones protein was identified as a monoclonal light chain of immunoglobulin [1].

Between 1928-1963, many substances including (hormones, enzymes, isoenzymes and proteins) were discovered and used as tumor markers useful in the diagnosis of individual tumors, but the general application of tumor marker for monitoring cancer in patients did not start until the discovery of α-fetoprotein (AFP)[2] in 1963 and carcinoembryonic antigen (CEA) in 1965[3]. The production of such markers during fetal development as well as in tumors, led to the term oncodevelopmental markers.

Monoclonal antibodies were developed in 1975[4]. to detect oncofetal antigens and antigens derived from tumor cell lines such as CA125, CA 15-3, and CA 19-9.

More recent advances in molecular techniques with the use of molecular probes and monoclonal antibodies to detect chromosome or protein alteration, including the study of oncogenes, suppressor genes, and genes involved in DNA repair, have led to the rapid understanding and use of tumor markers at molecular level [5]. Unlike earlier tumor markers, these new markers can be linked to specific biological processes related to the regulation of cell growth and tumor development including malignant transformation, proliferation, apoptotic cell death and metastasis.

## 1.2 Definition of tumor marker

A tumor marker is a substance that is present in or produced by tumor itself, or produced by the host in response to a tumor [6] that can be used to differentiate a tumor from normal tissue or to determine the presence of a tumor based on its measurement in the blood or secretions. Such a substance can be found in cells, tissue, or body fluids [7]. It can be measured qualitatively or quantitatively by chemical, immunological, or molecular biological methods to identify the presence of a cancer [8].

Morphologically, cancer tissue has been recognized by pathologists as resembling fetal tissue more closely than normal adult differentiated tissue. Tumors are graded according to their degree of differentiation: as being well differentiated, poorly differentiated, or anaplastic (without form). Few markers are specific for single individual tumor (tumor-specific markers); most are found with different tumors of the same tissue type (tumor-associated markers). They are present in higher quantities in cancer tissue or in blood of cancer patients than in benign tumors or in the blood of normal subjects.

Tumor markers have been categorized as enzymes, isoenzymes, hormones, and oncofetal antigens, carbohydrate epitopes recognized by monoclonal antibodies, receptors, oncogene product and genetic changes. There are only a handful of well-established tumor markers that are being used by physicians. Many other potential markers are still being under research. Now there are many studies that are trying to find new genes involved in signaling molecules or proteins that "tell" cells to proliferate, invade or metastasize [9].

## 1.3 Classification of Tumor Markers

Tumor markers may be classified into chemical and genetic tumor Markers. [10]

## 1.3.1 Chemical tumor markers

Table (1-1) Summarizes the classification of chemical tumor markers according to biochemical characteristics, and their associated malignancy.

### Table (1-1) Chemical tumor markers [11]

| Marker type | Associated malignancy | Example |
|---|---|---|
| Enzymes | Liver | Alcohol dehydrogenase |
| | Bone, Liver, Leukemia, Sarcoma | Alkaline phosphatase |
| | Ovarian, lung, trophoplastic, gastrointestinal, seminoma, hodgkin's | Alkaline phosphatase placental |
| | Pancreas, Various | Amylase |
| | Colon, Breast | Aryl Sulfatase B |
| | Colon, Bladder, Gastrointestinal, Various | Galactosyl Transferase |
| | Lung, (small-cell) neuroblastoma, Carcinoid, melanoma, pancreatic | Neuron-Specific enolase |
| | Prostate, Various (large bowel, lung, ovarian) | Prostate-specific antigen (PSA) Ribonuclease |
| | Colorectal, Breast, etc. | Telomerase |
| | Colon, Breast, Lung | Sialyl Transferenase |
| Hormone | Gushing's syndrome, Lung (small cell) | ACTH |
| | Lung (small cell) adrenal Cortex, deudonal | Antidiuretic hormone |
| | Medullary thyroid | Calcitonin |
| | Pituitary adenoma, Renal, Lung, Embryonal, choriocarinoma, Testicular (nonseminomatous) | hCG |
| | Torophoblastic, Gonads, Lungs, Breast | Human placental lactogen |
| | Liver, Renal, Breast, Lung, Various | Parathyroid hormone |
| | Pituitary adenoma, renal, lung, breast. | prolactin |
| Oncofetal Antigen | Hepato cellular, germ line (non seminoma) | -feto protein |
| | Colon | β- oncofetal antigen |
| | Liver | Carcino fetal ferritin |
| | Colorectal, Gastrointestinal, Pancreatic, | CEA |

13

| | | | |
|---|---|---|---|
| | | Lung, Breast | |
| | | pancreatic | Pancreatic oncofetal |
| | | Cervical, Lung, Skin, Head& neck (Squamous) | Squamous cell antigen |
| | | Colon,Gastrointestinal, Bladder | Tennesse antigen |
| | Mucin | Ovarian, endometrial, Lung | CA125 |
| | | Breast, ovarian | CA15-3 |
| | | Breast | CA27-29 |
| | | Breast, ovarian | MCA |
| | | Pancreatic, Ovarian, Gastrointestinal, Lung | Du-PAN-2 |
| | protein | Multiple Myeloma, β-Cell lymphpoma chronic | β- 2 microglobulin |
| | | Insulinoma | C-peptide |
| | | Liver, Lung, Leukemia | Ferritin |
| | | multiple myeloma, Lymphomas | Immunoglobulin |
| | | Pancreatic, Stomach | Pancreas associated antigen |
| | | Trophoplastic, Germ cell | Pregnancy specific protein |
| | | Hepatocellular | Prothrombin precursor |
| | | Ovarian | Tumor associated trypsin inhibitor |
| Blood Group Related antigen | | Pancreatic, Hepatic, Gastrointestinal, | CA19-9 |
| | | Pancreatic, Gastrointestinal, Ovarian | CA19-5 |
| | | Colon, Pancreatic Gastrointestinal, | CA50 |
| | | Ovarian, Breast, Colon ,Gastrointestinal, | CA72,4,CA242 |
| Others | | Breast | Estrogen&progesterone receptors |
| | | Brain, Various | polyamine |
| | | Bone Metastasis (Breast), (multiple myeloma) | Hydroxy proline |
| | | Neuroblastoma, pheochromocytoma | Catecholamine metabolites |
| | | Gastroitestinal, Lung, Rhenmatoid | Lipid-associated sialic acid |

## 1.3.2 Genetic tumor markers

A simple definition of cancer is "a relatively autonomous growth of tissue". Understanding the cause of autonomous growth would clearly facilitate the search for care. Advance in molecular genetics have provided a better understanding of the genesis of human cancer. The proliferation of normal cell is thought to be regulated by growth – promoting oncogenes and counterbalanced by growth – constraining tumor suppressor genes. The development of cancer appears to involve the activation or altered expression of oncogenes, and the loss or inactivation of tumor suppressor genes.[11]

Oncogenes (cell activation genes) are derived from proto- oncogenes (normal celluler genes) which may be activated by dominant mutations. The type of mutation might be point mutation, insertion, deletion, translocation, or inversion. Most oncogenes are associated with haematological malignancies such as leukemia and to lesser extent, solid tumors. [12]

Suppressor genes (genes involved in the recognition and repair of damaged DNA) have mostly been isolated from solid tumors. p53 tumor suppressor gene is the most frequently mutated gene in human cancer, indicating its important role in the conservation of normal cell cycle progression [13]. One of p53'S essential roles is to arrest the cells in G1 after genotoxic damage, to allow DNA repair prior to DNA replication and cell division. In response to massive DNA damage, p53 triggers the apoptotic cell death pathway [14]. The loss of function of this gene may result in the inability of DNA repair process and may lead to the development of tumorgensis [15].

The exciting promise of using the detection of oncogenes and suppressor genes, for diagnosis, determining the prognosis, and predicting the response to

15

the chemotherapy remains to be realized. Oncogens detection remains an experimental approach to human cancer.

## 1.4 Ideal Tumor Marker

An ideal tumor marker should be specific for a given type of cancer and sensitive enough to detect small tumors for early diagnosis or during screening, provide an estimation of tumor burden, and serve for monitoring effects of therapy and detecting recurrence of tumors [7]. It can be measured qualitatively or quantatively by chemical, immunological, or molecular biological methods to identify the presence of a cancer [8].

## 1.5 Clinical Applications of Tumor Markers

Clinical applications of tumor markers depend on specificity and sensitivity. Specificity refers to the detection of specific tumors by specific markers. Sensitivity, in this instance, has to do with detecting all patients with the specific tumor.

The potential uses of tumor markers are [16]:

- Screening in general population.
- Differential diagnosis of symptomatic patients.
- Clinical staging of cancer
- Estimating tumor volume.
- Prognostic indicator for disease progression
- Evaluating success of treatment.
- Detecting response to therapy
- Radioimmunolocalization of tumor marker.
- Determining direction for immunotherapy.

In general tumor markers may be used for diagnosis, prognosis, and monitoring effect of therapy, as well as a target for tumor localization and therapy. Monitoring treatment with tumor markers is an accepted application and generally indicates successful treatment; such monitoring is seen after both invasive and noninvasive treatment.

The use of tumor markers for screening the presence of cancer in asymptomatic individuals in a general population has been limited because most tumor markers are present in normal, benign and cancer tissues and are not specific enough to be used for screening cancer [17]. However if the incidence of cancer is high among certain population, screening is feasible [18]. Potential uses of some tumor markers are summarized in table (1-2).

**Table (1-2) Clinical Uses of some tumor markers [19]**

| Tumor marker | Biochemical properties | Molecular weight | Primary clinical applications |
|---|---|---|---|
| Alpha–fetoprotein ( AFP) | Glycoprotein, 4% carbohydrate; considerable homology with albumin | ~70kDa | Diagnosis and monitoring of primary hepatocellular carcinoma and germ cell tumors. Prognosis of germ cell tumors. |
| Cancer antigen 125 (CA125) | Mucin identified by monoclonal antibodies | ~ 200kDa | Monitoring ovarian carcinoma. Prognosis after chemotherapy |
| Cancer antigen 15-3 (CA15.3) | Mucin identified by monoclonal antibodies | >250 kDa | Monitoring breast cancer |
| Cancer antigen 72.4 (CA 72.4) | Glycoprotein identified by monoclonal antibodies | ~48 kDa | Monitoring gastric carcinoma |
| Cancer antigen 19-9 (CA19-9) | Glycolipid carring the Lewis a blood group determinate | ~1,000 kDa | Monitoring pancreatic carcinoma |
| Carcinoembri - yonic antigen (CEA) | Family of glycoproteins, 45%-60% carbohydrate | ~180 kDa | Monitoring gastrointestinal and other Adenocarcinomas |
| Estrogen receptor | Nuclear transcription | 65 kDa | Predicting response to endocrine therapy in breast cancer |

17

| Human chorionic gonadotrophin (hCG) | Glycoprotein hormone consisting of tow non-covalently bound subunits ( α and β) | ~36 kDa | Diagnosis and monitoring non- seminomatous germ cell tumors, Prognosis of germ cell tumors. |
|---|---|---|---|
| Placental alkaline phosphatase (PLAP) | Heat – stable isoenzyme of alkaline phosphatase | ~86 kDa | Monitoring of germ cell tumors (seminomas) |
| Progesterone receptor | Nuclear transcription factor | A from : 94 kDa B from :120 kDa | Predicting response to endocrine therapy in breast cancer. |
| Prostate specific antigen (PSA) | Glycoprotein serine protease | ~ 36 kDa | Diagnosis, screening and Monitoring prostatic carcinoma |
| Tissue polypeptide antigen (TPA) | Fragments of cytokeratin 8,18 and 19 | ~22 kDa | Monitoring bladder and lung carcinoma |
| Tissue polypeptide specific antigen (TPS) | Fragments of cytokeratin 18 | ~22 kDa | Monitoring metastatic breast carcinoma |

## 1.6 Tumors of Ovary

## Classification

The pathological conditions of the ovary may be classified as [20, 21]:

### *1.6.1 Non-neoplastic Functional Cysts*

Follicular cyst

Corpus luteum cyst.

Theca lutein and granulose lutein cysts.

Polycystic ovarian disease.

Endometriomatous cysts.

### *1.6.2 Ovarian Neoplasm's.*

The classification of ovarian neoplasm's given in table (1-3), is a simplified version of world health organization (WHO) histological classification, which separates ovarian neoplasm's according to the most probable tissue of origin. [22]

**Table (1-3) Derivation of various ovarian neoplasms' and some data on their frequency and age distribution [22]**

| Origin | Surface Epithelial cells (surface Epithelial – stromal cell tumor ) | Germ cell | sex cord - stroma | Metastasis to ovaries |
|---|---|---|---|---|
| overall frequency | 65-70% | 15-20% | 5-10% | 5% |
| Proportion of malignant ovarian tumors | 90% | 3-5% | 2-3% | 5% |
| Age group affected | 20 + years | 0-25 + years | All ages | Variable |
| Types | - Serous tumor<br>- Mucinous tumor<br>- Endometrioid Tumor<br>- Clear cell Tumor<br>- Transitional cell Tumor<br>- Undifferentiated carcinoma | - Teratoma<br>- Dysgerminoma<br>- Endodermal sinus tumor<br>- Chorio carcinoma | - Fibroma<br>- Granulos-theca cell tumor<br>- Sertoli – leydig cell tumor | |

## 1.7 Epithelial Ovarian Tumors

The most common group of ovarian neoplasm originates from the coelomic mesothelium that covers the ovary, which after neoplastic transformation seems to retain the capacity to recapitulate the epithelial components of the mullerian ducts. According to the (WHO) classification of ovarian tumors, surface epithelial-stromal tumors can be divided into serous tumors, mucinous tumors, endometrioid tumors, clear cell tumors, transitional tumors and undifferentiated carcinoma [23, 24,25] table (1-3).

Serous tumor is the most frequent of the ovarian tumors and the most common epithelial ovarian carcinoma, accounting for 40-50% of all such tumors. [26]

According to histopathological classification of ovarian tumors, serous tumors can be classified into. [27]

## Benign

- Cystadenoma and papillary cystadenoma.
- Surface papilloma
- Adenfibroma and cyst adenofibroma .

## Borderline malignancy (of low malignant potential)

- Cystic tumors and papillary cystic tumor.
- Surface papillary tumor.
- Adenofibroma and cystadenofibroma.

## Malignant.

- Adenocarcinoma, papillary adenocarcinoma.
- Cystadenocarcinoma.
- Surface papillary adenocarcinoma.
- Adenocarcinofibroma & cystadenocarcinofibroma.

Evidence is lacking about whether ovarian carcinoma may go through a borderline phase during its development and whether borderline tumors always shift into invasive ovarian carcinoma [28-29].

In the overall spectrum of serous tumors, about 60% are benign, 15% of low malignant potential, and 25% malignant [22]. Benign serous cystadenomas occur slightly more often than benign mucinous tumors, but in their malignant form serous cyst adenocarcinomas are three to four times more common than mucinous cystadenocarcinomas. Serous and mucinous borderline tumors are seen but other types of epithelial tumors of borderline malignancy, such as the variant of Brenner and endometrioid tumors are rare [30].

## *1.7.1 Incidence*

Ovarian cancer is the leading cause of death from gynecological malignancies [31] worldwide, the highest incidence of disease is found in America and northern Europe, the lowest incidence is found in Asia and Latin Amarica[32]. In United States approximately 23000 cases occur annually leading to 13900 death each year [33], rate among blacks are lower than among whites, but rates for women of Chinese and Japanese are higher than rates in their countries of origin [32]. In Iraq, ovarian cancer forms 38% of all gynecological malignancies, it was the seventh most common cancer among females with an incidence of 0.8 per 100,000 woman [34]. It is the most common cancer to occur at an advanced stage [35].

## 1.7.2 Etiology and Risk Factors of ovarian cancer

Although the exact etiology of ovarian cancer is unknown, there are many risk factors which have been associated with the developing of ovarian cancer, such as Genetic, environmental, hormonal, and nutritional.

### 1.7.2.1 Genetic factors

The strongest known risk factor for ovarian cancer is family history which is present in about 10-15 % of women who develop the disease [36]. Family history in one relative increased the lifetime probability of ovarian cancer in a 35- years old women from 1.5 to 5%; the probability increased to 7% if she had two relatives with the disease.

The rare familial ovarian cancer syndromes are accounting for less than 1% of ovarian cancer cases. The most common hereditary syndrome is the breast-ovarian cancer syndrome. Most of these families have germ- line mutation in one of the breast cancer susceptibility genes, BRCA-1 [37] or BRCA-2. [38]

### 1.7.2.2 Increasing Ages

The incidence of ovarian cancer increases with age, the highest proportion of cases is diagnosed in women 50 to 59 of age. In women 50 to 75 years of age, the annual incidence is 50 per 100,000 (adjust for prior oophorectomy), approximately twice the rate in young women [39, 40], while benign tumors occur mostly in young women between the ages of 20-40 years old. [22]

### 1.7.2.3 Reproductive factors

Several potentially modifiable reproductive factors appear to reduce the risk of ovarian cancer

- Pregnancy reduces the odds of ovarian cancer by 25 to 50 percent [41,42] the decrease in risk is associated with an increasing number of pregnancies.

- Use of the oral contraceptive pill is associated with a 35 % reduction in the risk of ovarian cancer,  increasing the duration of use is associated with decreasing risk, [41,42] ten years of use by women with a positive family history can reduce their risk to a level below that for women with no family history who never used oral contraceptives. [43]

- Breast feeding is associated with a more modest effect on risk, reducing the odds of ovarian cancer by 20 % [41,42].

Certain gynecological surgical procedures are also associated with a lower likelihood of ovarian cancer. The risk of ovarian cancer is reduced by about 15% after tubal ligation or hysterectomy with ovarian preservation [41,42].The protective mechanism of these procedures may relate to impairment of ovarian function, causing an ovulation, or protection from exposure to exogenous carcinogens that enter the peritoneal cavity through the vagina. Talcum powder may be one of the carcinogen materials,  studies have shown that woman who use talcum powder as part of their perineal hygiene are at increased risk. [43] Talc is found in soap powders, and deodorants, and is used in packing of condoms and [30, 44, 45] contraceptive diaphragms, talc might then migrate through vagina to reach the ovaries.

Infertility may increase the risk of ovarian cancer [46, 47]. An increased risk of ovarian cancer has also been reported with infertility treatment, particularly prolonged use of clomiphene citrate. [48]

## *1.7.3 Staging*

Staging of ovarian cancer is based on the finding at the time of surgery and pathological review. Because of the clinically occult spread, surgery is mandatory. The clinical staging of cancer is intended to provide means by which information related to the progress of the disease, the methods and success of treatment modalities is obtained.

Ovarian malignancies are staged according to the international federation of gynecology and obstetrics (FIGO), basing on the finding of surgical exploration[49]   see table (1-4).

.

**Table (1-4) Staging of ovarian cancer according to FIGO.** [ 49 ]

| Stage I |
| --- |
| In stage I, cancer is found in one or both of the ovaries. Stage I is divided into |
| Stage IA: Cancer is found in a single ovary. |
| Stage IB: Cancer is found in both ovaries |
| Stage IC: cancer is found in one or both ovaries and one of the |

following is true:

Cancer is found on the outside surface of one or both ovaries.

The tumor has ruptured the ovary wall.

Cancer cells are found in fluid from the peritoneal cavity (the body cavity that contains most of the organs in the abdomen). The fluid may already be in the peritoneal cavity or it may be added by the doctor to wash the peritoneum (tissue lining the peritoneal cavity).

## Stage II

In stage II, cancer is found in one or both ovaries and has spread into other areas of pelvis. Stage II is divided into

Stage IIA: Cancer has spread to the uterus and / or the fallopian tubes.

Stage IIB: Cancer has spread to other tissue within the pelvis.

Stage IIC: cancer has spread to the uterus and/or fallopian tubes and /or other tissue within pelvis and one of the following is true:

Cancer is found on the outside surface of one or both ovaries.

The tumor has ruptured the ovary wall.

Cancer cells are found in fluid from the peritoneal cavity.

## Stage III

In stage III, cancer is found in one or both ovaries and has spread to other parts of the abdomen. Cancer has spread to the surface of the liver. Stage III is divided into:

Stage IIIA: The tumor is found only in the pelvis, but cancer cells have spread to the surface of the peritoneum.

Stage IIIB: Cancer has spread to the peritoneum but is not larger than 2 centimeters (less than 1 inch) in diameter.

Stage IIIC: Cancer has spread to the peritoneum and is larger than 2 centimeters in diameter and/or has spread to lymph nodes in the abdomen.

## STAGE IV

In stage IV, cancer is found in one or both ovaries and has metastasized (spread) beyond the abdomen to other parts of the body .Cancer is found in the tissues of the liver.

## 1.7.4 Diagnosis

Ovarian tumors are occasionally detected on pelvis examination, although early stage tumors are rarely found due to deep anatomic location of the ovary. Thus tumors detected by pelvic examination are usually at an advanced stage and associated with poor prognosis [50], more than 56-70% of the patients are diagnosed as having stage III or IV disease. [51]

Tests that examine the ovaries, pelvic area, blood, and ovarian tissue are used to help diagnose ovarian cancer.

These include the following. [49]

- **Pelvic exam**: A procedure to check the uterus, Vagina, ovaries, fallopian tubes, bladder, and rectum to find any abnormality in their shape or size.

- **Ultrasound test**: Transvaginal sonography (TVS) is capable of detecting more than 95% of stage I ovarian cancers, it also detects large numbers of patients with benign disease who subsequently undergo surgery to rule out malignancy, the predictive value was only 1.5%. [52]

- **CA125 Test**: A blood test used to measure the level of CA125 [53], a substance sometimes found in an increased amount in the blood, other body fluids or tissues.

- **Computerized axial tomography (CAT scan)**: A series of detailed pictures of areas inside the body, taken from different angles. This method fails to differentiate benign disease from stage I disease. [54, 55]

- **Intravenous pyelogram**: A series of x-ray of the kidneys, Ureters, and bladder to help determine if cancer has spread outside the ovaries.

- **Biopsy**: Removal of tissue for examination under microscope.

### 1.7.5 Treatment

There are treatments for all patients with ovarian epithelial cancer. Some treatment are standard, and some are being tested in clinical trials. A treatment clinical trial is a research study meant to help improve current treatments or [49] obtain information on new treatments for patients with cancer.

**1.7.5.1 Standard treatment: three kinds of standard treatment** are used, these include the following:

#### ⟶ Surgery

Most patients have surgery to remove as much of the tumor as possible. Different types of surgery may include

- Hysterectomy (removal of the ovaries, fallopian tubes, and uterus).
- Unilateral Salpingo-oophorectomy (removal of one ovary and one fallopian tube).
- Bilatera salpingo-oophorectomy (removal of both ovaries and both fallopian tubes).
- Omentectomy (partial removal of the lining of the abdominal cavity).
- Lymph node biopsy (removal of the lymphnodes for examination under a microscope to check for cancer cells).

#### ⟶ Radiation therapy

Radiation therapy is the use of x- ray or other types of radiation to kill cancer cells and shrink tumors. Radiation therapy may use external or internal radiation.[49]

#### ⟶ Chemotherapy

Chemotherapy is the use of drugs to kill cancer cells. Chemotherapy may be taken orally or injected into a vein or muscle. Another way to give chemotherapy is intraperitoneal chemotherapy. With this method, most of the drug remains in the abdomen; this technique is effective in advanced disease with minimal residual disease. [44]

⟶ Biological therapy

- High-dose chemotherapy with bone marrow transplantation.

Biological therapy is the treatment to stimulate the ability of the immune system to fight cancer. Substances made by the body or made in laboratory are used to boost, direct, restore the body's natural defense against disease.

Biological therapy is sometimes called biological response modifier (BRM) therapy or immunotherapy. [49]

Within the last few years, different immunotherapeutic strategies based on immunization with tumor specific antibody constructs or immunogenic peptides have been developed [56, 57]. Alternative concepts include the application of genetically modified tumor cells for the expressions of cytokines or consitimulatory molecules as well as dentritic cells for the effective presentation of immunogenic peptides in the extent of MHC and the activation of cellular immune responses, [57] in this context, ovarian cancer cells express a mutated form of the P53 [58, 59, 60] and/or BRCA1 [61, 62] (Tumor suppressor gene).

Ovarian tumors also over express the CA125 [63] and $DF_3$/MUCI [64] carcinoma-associated antigens. In addition, these tumors over express the $HER_2$/neu (c-erbB$_2$) [65, 66] thus certain targets for immunotherapy of cancer are already known and others although remain undefined presumably exist.

**Stage I**: Treatment of stage I may include hysterectomy, unilateral or bilateral salping-oophoretomy and omentectomy. It also may include radiation therapy, chemotherapy and clinical trail conservative unilateral salpingo oophorectomy is adequate. It appears that ovarian preservation and women's fertility is safe and reasonable in women of reproductive age. [44, 67]

**Stage II**: treatment of stage II may include surgery to remove the tumor (hysterectomy, bilateral salpingo-oophorectomy, and omentectomy). Combination with chemotherapy or radiation therapy gives approximately 85-90% five years survival. [44, 68]

**Stage III**: treatment of stage III may include surgery to remove the tumor (hysterectomy, bilateral salping - oophorectomy, and omentectomy) followed by chemotherapy or chemotherapy followed by second look surgery.

**Stage IV**: Treatment of stage IV ovarian epithelial cancer is combination chemotherapy with or without surgery to reduce the size of the tumor. Extensive surgery is often insufficient to eliminate the intra-abdominal tumor and response to chemotherapy is only partial in many of these patients[44].

## 1.8 Ovarian Tumor Markers

Ninety percent of ovarian malignancies are epithelial. [25, 69] There are quantitative and qualitative changes in numerous circulating substances which have been associated with epithelial ovarian cancer. These may reflect an alteration in ovarian function, surface molecular structure or general responses to malignancy.

Expression of specific antigens associated with epithelial ovarian cancer is useful for establishing a diagnosis, classification and providing prognostic information. Monitoring the appropriate antigen titers is very useful in the identification of occult metastasis, monitoring of therapeutic response and detection of asymptomatic recurrence at an early stage. [70, 71, 72]

The antigens defined on ovarian tumors are regarded as tumor-associated rather than tumor-specific [73]. Several tumor markers have been investigated for one or more clinical use in ovarian cancer as shown in table (1-5).

**Table (1-5): Tumor markers that have been investigated in ovarian cancer** [10].

| Marker type | Example |
|---|---|
| Hormones | Progesterone [74], Estrogene , urinary gonodotrophin fragment. |
| Enzymes | Placental alkaline phosphatase[75,76] creatine |

28

| | | kinase, amylase glactosyl transferase, ribonuclease |
|---|---|---|
| Oncofetal antigens | | Tissue polypeptide antigen[77] alpha-fetoprotein ,carcinoembryonic antigen( CEA) [78] . |
| Carbohydrate markers | | A:Mucin tumor markers: CA125, CA15-3 B. blood group antigens related cancer marker. CA72-4 [79] . |
| Genetic markers | Oncogene products | C- erb B-2 amplification HER-2/neu [80] |
| | Tumor suppressor genes | BRCA1[37] ,BRCA2 [38] |

There are several tumor markers that correlate with the incidence of ovarian cancer, but the most important markers are:

### *1.8.1 CEA*

Carcinoembryonic antigen (CEA), one of the onco-fetal proteins, is a cell surface glycoprotein, with a high molecular mass of 150-300 kDa. It is normally expressed in the early embryonic development and tends to disappear with the onset of differentiation of fetal tissue into adult ones. CEA has been studied extensively in ovarian cancer and has been reported to be elevated in 30-65% of epithelial tumors, mainly in patients with advanced stage disease. This antigen has been shown not to correlate well with status of disease. [81]

### *1.8.2 CA 19.9*

CA19.9 is a carbohydrate antigen that is measured by a monoclonal antibody and can be found elevated in only 17-25% of patients with epithelial malignancies [82,83].

### *1.8.3 CA 15-3*

CA15-3 is a mucin-like membrane glycoprotein recognized by a pair of monoclonal antibodies: the murine antibody DF-3 and 115D8 [64]. Distinct epitopes of this high molecular weight antigen (300-400kDa) [84] is the carbohydrate side chain which accounts for about 50% of its structure [85].

CA15.3 is found in adenocarcinoma of breast, lungs, ovaries and pancreas[86, 87]. Its level is elevated in 64% of ovarian cancer, it is most useful tumor marker for breast cancer [88].

### 1.8.4 IL.6 and IL.10

The interleukins, IL-6 and IL-10, have been shown to be present in very high levels in the ascites and serum of women with advanced stage epithelial cancers [89, 90]. IL-6 correlates well with the stage and status of disease, but is elevated in only about 66% of patients, and its complementarily with CA125 is only modest.

### 1.8.5 M-CSF

Macrophage colony stimulating factor (M-CSF) has been found to be measurable in the serum of 68% of patients with clinically detectable disease[91,92].

Interestingly, some complementarities with CA125 has been documented. Patients with clinically evident tumor and a negative CA125 ($< 35$ u.ml$^{-1}$), 56% had an elevated M-CSF serum level [92].

### 1.8.6 LSA

Lipid-associated sialic acid (LSA) can be measured in the sera of about 60% of patients with ovarian cancer, mostly those with advanced stage disease. A combination of LSA and CA125 improve sensitivity for detection of advanced disease but does not improve specificity [82].

### 1.8.7 OVXI

OVXI antigen is a high molecular weight mucin. The antigenic determinate of OVXI antigen was raised by immunization of mice with human ovarian cancer cell line. OVXI antigen is elevated in 67%of patients with clinically evident ovarian cancer who are CA125 negative. The OVXI is elevated however, in only 45%of patients with ovarian cancer. In patients with residual disease at second-look surgery and a negative CA125, 27% had an elevated OVXI level in the serum.[93]

## 1.8.8 NB 70 K

NB70K antigen appears to be present in most major types of epithelial ovarian cancer, with an apparent molecular weight of 70 kDa [94]. Among sera samples from ovarian cancer patients ,elevated NB70K levels were found in 87% of samples that contained elevated CA125 levels. No quantitative correlation was found, however, between levels of NB70K and CA125.[94,95]

## 1.8.9 TAG 72

Tumors associated glycoprotein (TAG-72) level is elevated in 50% of ovarian carcinoma cases and only in 4% of benign disease cases with the highest level of expression in mucinous cystadenocarcinoma and its measurement may be useful as a confirmatory tumor marker for the presence of ovarian cancer in those patients with elevated CA125 serum levels. Combined TAG-72 and CA-125 test increase the sensitivity for the detection of primary ovarian cancer to 73% with no significant change in specificity [96].

## 1.9 Cancer Antigen 125(CA125)

### 1.9.1 Biochemistry

Cancer antigen 125(CA125) was first defined by monoclonal antibody (OC125) more than 20 years ago [53]. It was associated with a family of high molecular weight glycoproteins, that differed from classical mucins by means of carbohydrate conversion(less than 50%) and presence of both N and O linked carbohydrate residues [97,98]. Size exclusion chromatography of native CA125 antigen material from body fluids results in at least two broad peaks of antigen reactivity with approximate relative molecular mass of 200 and >1000 kDa,[99] but lower molecular weight species have also been reported .

Chemical study has revealed sensitivity of CA125 to proteases; low pH, high temperature, and high ionic strength, properties consisted of conformational peptide determinant. However, CA125 activity can also be destroyed with relatively high concentration of periodate and blocked with different lectins, suggesting a close association with carbohydrate. [97]

### *1.9.2 Structure*

Although CA125 biochemical nature has long been elusive, its primary structure was established four years ago, indicating that it is trans-membrane protein with a short intracellular and a giant extracellular domain, the latter is with 22,097 amino acid residues. The extracellular part is composed of an amino-terminal part spanning 12,070 residues [100], followed by more than 60 tandem repeats of 156 amino acid motif and 229-residues linker to trans-membrane domain [101]. Both the amino terminal part and the repeat domains are rich in serine and thereonine residues and are highly glycosylated. The carbohydrate content was estimated to be 24-28%, with O-linked and N-linked glycans [102]. Because highly O-glycosylated repeats are the landmark of the mucin family of glycoproteins, CA125 was also named MUC 16 to reflect the nature of CA125 as a new member of protein family of mucins [103]. The mucin-like repeats contain a domain that was reported to be susceptible to proteolytic cleavage [104]. An additional potential proteolytic cleavage site in CA125 was reported to be located immediately membrane-proximal [101]. Figure (1-1) represents a proposed structure of CA125 antigen.

## 1.9.3 FUNCTION OF CA125

Although primary structure of CA125 has been elucidated, a functional

role for this molecule in physiological context or in cancer remains unknown. However, a number of publications have pointed out several properties of CA125 that may be of relevance for its biological function.

First, because of its expression in embryonic membranes and adult derivatives of the fetal periderm, CA125 has been suggested to play a role as a lubricant, preventing adhesion of membranes [105]. Anti adhesive properties have also been assigned to other mucins [106].

Second, close analysis of glycans present on CA125 protein revealed the presence of several glycan structures that have been implicated in immune suppression [102] raising the possibility that CA125 might help protect the embryo from maternal immune rejection and play an immunovasive role in ovarian cancer. Furthermore mucin can bind to various sugar-binding molecules, such as galectins [107]. CA125 was found to be a novel counter receptor for galectin-1[108]. The known cellular responses to the cell-surface recruitment of galectin-1 include a change in proliferation activity, regulation of cell survival and regulation of cell adhesion. Depending both on the cellular context and its local concentration, galectin-1 exerts both inhibitory and stimulatory effects on these processes [109].

Third Gaetje et al. found that CA125 from human peritoneal fluid was shown to enhance the invasiveness of a benign endometriolic cell line EEC145, but it did not affect the invasiveness of a variety of non-endometrioid cell lines, raising the possibility that CA125 plays a role in endometriosis [110].

Recently, Rump et al. in 2004 [111] have demonstrated that mesothelin (a glycoprotein which is present in peritoneal fluid of ovarian cancer patients) [112] is a novel CA125-binding protein and they (CA125 and mesothelin) are co-expressed in advanced grade ovarian adenocarcinoma, which indicates that CA125 might contribute to the metastasis of ovarian cancer to the peritoneum by initiating cell attachment to the mesothelial epithelium via binding to mesothelin.[111]

## 1.9.4 Expression

CA125 is derived from celomic epithelium (pleura, peritoneum and pericardium), amnion and Mullerian duct during embryonic development. Trace amounts of CA125 are found in adult tissues derived from the epithelial lining of the pleura, peritoneum, pericardium, fallopian tube, endometrium and endocervix [98, 113].

Relative to the expression levels of CA125 found in normal tissues, CA125 is often overexpressed from epithelial ovarian cancer tissue and other tumors of non-gynecological malignances.[53, 114]

Although CA125 is expressed both by normal and tumor cells, cell surface expression and release of soluble proteolytic fragments of CA125 into the extracellular space [115] appear to be associated with the conversion from benign to cancer cells [116].

## 1.9.5 Developing The CA 125 Assay

The development of an assay for the CA125 tumor marker grows out of attempts of Bast et al [117] which aimed to obtain monoclonal antibodies for serotherapy of patients in ovarian cancer. In this attempt, mice were repeatedly injected with a human ovarian cancer cell line and hybridomas were prepared from immune spleen cells and the P3NS-1 plasmacytoma. From these hybridomas, clones were isolated based on the production of antibodies that bound to the ovarian cancer cell line used for immunization, but not to a B lymphocyte cell line developed for the tumor cell donor. The one hundred twenty-fifth promising clone produced an IgG1 antibody of the desired specificity and was designated as OC (ovarian cancer). [117]

Using immunohistochemical analysis, the OC 125 antibody was found to bind to antigen expressed by approximately 80% of epithelial ovarian cancer as well as by other gynecologic, breast, lung, and colon carcinomas: this antigen was designated as CA (cancer antigen) 125.[117]

The first monoclonal antibody radioimmunoassay for monitoring epithelial ovarian cancer, using OC125 antibody, was reported in 1983 by Bast *et al.* [53]. After that several different formats for the assay have been developed using radiolabeled or enzyme-labeled OC 125 as a probe. Over the last decade, a number of monoclonal antibodies have been developed that react with one or two distinct epitopes on molecules expressing CA125. One of these antibodies, M11, has permitted the development of a second generation assay, CA125II, in which M11 is used to trap antigen, followed by OC125 to detect antigen that has been captured on a solid phase [118].

## *1.9.6 Clinical Applications of CA125*

The best available marker for epithelial ovarian cancer is CA125. The normal range most frequently quoted for CA125 is 0-35 u.ml$^{-1}$, although 99% of apparently healthy post-menopausal women have levels below 20 u.ml$^{-1}$. In apparently healthy pre-menopausal women, levels of 100u/ml or higher can occur during menses.[75] Elevation was also observed with the first trimester of normal pregnancy.[119] Although elevated CA125 levels are found in approximately 80% of all patients with epithelial ovarian cancer, high levels are found in only about 50% of stage I disease. [120]

This lack of sensitivity for early disease, and the fact that CA125 can be elevated in multiple benign disease such as endometriosis [121,75] , limits the use of CA125 for the diagnosis of early epithelial ovarian cancer. Further limitation is that CA125 can be elevated in adenocarcinomas other than ovarian cancer. Although CA125 can also be elevated in germ cell tumors of ovary [120], the markers of choice for this type of ovarian cancer are αβ-fetoprotein (AFP) and human chorionic gonadotrophin (hCG) and its β-subunite (β-hCG ).

### 1.9.6.1 Screening

The lack of early symptoms means that approximately 70% of the patients with ovarian cancer present with advanced disease. While the overall five- year relative survival rate is of the order of 30%, the survival rate for stage III and IV disease combined is only 10% [122]. In contrast, a five year survival of 90% may be achieved for patients with early stage disease confined to the ovary [123].

As a screening test, the main problems with CA125 are lack of sensitivity for early stage disease (only about 50% of patients with stage I have elevated levels) and lake of specificity. Thus, a single measurement of CA125 is not an adequate screening tool for ovarian cancer [124]. The rate of change in CA125 levels over time appears to be more specific screening method. In one study, the specificity reached 99.9% after redefining a positive test as CA125 concentration greater than 35 u.ml$^{-1}$ was doubles within six months.[124]

CA125, however, in combination with transvaginal ultrasound may have the role in the early detection of ovarian cancer. This screening strategy achieved a specificity of 100%, and an apparent sensitivity of 81.7% [125]. Other reports have found that the use of tumor markers complementary to CA125 (eg. OVX1) is useful to achieve a specificity of 99.9% and an apparent sensitivity of 80%. Measurement of complementary serum markers can be used as primary screening technique followed by transvaginal ultrasongraphy. This could provide  cost-effective means of early detection and could significantly decrease the probability of surgical intervention for false-positive test results.[126]

### 1.9.6.2 Diagnosis

The diagnosis of ovarian cancer is usually carried out by surgery followed by histopathology. However, pre-operative serum levels of CA125 especially in post-menopausal women, may be useful in the differential diagnosis of benign and malignant pelvic masses. Among post menopausal patients with a pelvic mass, a CA125 level greater than 65 u.ml$^{-1}$ has distinguished malignant disease with greater than 90% accuracy. The accuracy of CA125 (cut-off concentration 35 u.ml$^{-1}$) in differentiating between benign and malignant masses was 77%, which was almost identical to accuracy achieved with pelvic examination and Ultrasound (76% and 74% respectively) [127]. The combination of Ultrasound, CA125 and pelvic examination, however, improved the accuracy. Significantly, no cancer was found in any subject in which all three tests were negative. [127]

### 1.9.6.3 Prognosis (chance of recovery)

The traditional prognostic factors for ovarian cancer include tumor stage, grade, histological type and size of residual tumor after primary debulking (cytoreductive surgery). However multiple studies have shown that CA125 levels after either 1,2 or 3 courses of chemotherapy is one of the strongest available indicators of disease outcome [128].

A prolonged half life for CA125, or decrease in CA125 concentration of less than 7 folds of pretreatment concentration, during the early month of treatment, has been suggested to be an indicator for a poor outcome. CA125 concentration >70 u.ml$^{-1}$ before the third course of chemotherapy was the single most important factor for predicting disease progression at twelve months [128].

### 1.9.6.4 Monitoring

The most important application of CA125 is in the monitoring of patients with epithelial ovarian cancer. Serial CA125 levels can pre-clinically detect recurrent disease with lead times of 1-17 months (median 3-4 months)[75,120], Doubling of initially elevated CA125 levels has been associated with disease progression in more than 90% of cases [75]. Furthermore, longitudinal monitoring with this marker has the potential to detect recurrent disease earlier and more cost-effectively than radiological procedures [128]. While early detection of recurrent disease may lead to altered patient management, no study has yet shown that this leads to enhanced survival. The use of CA125 and other markers for monitoring, will attain greater importance, as more effective treatment becomes available for previously - treated ovarian cancer.

### 1.9.6.5 Treatment

Induction of specific immune responses by vaccination with murine monoclonal anti-idiotypic antibody (Anti-CA125), which imitates the tumor-associated antigen CA125, has a positive influence on the survival of patients with recurrent ovarian carcinoma. Patients subjected to this immunotherapy technique showed increased concentration of human anti mouse antibodies. Specific anti-anti-idiotypic antibodies, as a marker for induced immunity, were detected in 66% of treated patients. Survival of patients with a positive immune response was 19.9±13.1 months in contrast with 5.3±4.3 months in those patients without detectable Anti CA125-immunity.[129]

The explanation of this specific immuno response caused by vaccination with anti-ioditypic antibody is that the variable antigen binding regions of antibodies (Ab1) contain idiotypic determinants that are immunogenic and induce the formation of so-called anti-idiotypic antibodies (Ab2), some of these antibodies are able to functionally mimic the three dimensional structure of original antigen, thus selective immunization with Ab2 could induce specific immune reaction directed against the original antigen. [129]

Antitumor vaccines were also developed by fusions of tumor associated antigens with dendritic cells. Human ovarian carcinomas express the CA125, $HER_2$/neu, and MUC1 Tumor associated antigens which are potential targets for the induction of active specific immunotherapy. Fusions of ovarian cancer cells to dentritic cells resulted in the formation of heterokaryons that express the CA125 antigen and dendritic cells-derived costimulatory and adhesion molecules. The fusion cells have been shown to stimulate proliferation of autologous T cell that induce cytolytic T-cell activity and lysis of autologous tumor cells. [130]

# Chapter Two

# Preliminary studies for the binding of $^{125}$I- antiCA125 antibody to the CA125 in Human Sera and homogenates of benign and malignant ovarian tumors.

## Abstract

Measurements of the two biochemical tumor markers CA125 and CEA were carried out in serum samples obtained from 30 healthy donors, 34 ovarian cancer patients (20 post-menopausal patients with malignant ovarian tumors (OI )and 14 pre-menopausal patients with ovarian tumors (OII) and 24 ovarian benign patients (OIII) using Immunoradiometric assay (IRMA) technique.

Mean values of CA125 in pre-menopausal patients with benign and malignant ovarian tumors and CEA in pre-menopausal patients with ovarian cancer were found to be significantly higher ($P<0.05$) than their corresponding values in sera of healthy control, while insignificant differences ($P>0.05$) between CEA values in patients with benign tumor and healthy control were found. In comparison between post-menopausal and pre-menopausal patients with ovarian cancer tumor, values of CA125 and CEA in post-menopausal were significantly higher ($P<0.05$) than those found in pre-menopausal group. The tumor marker CA125 shows the best sensitivity 80% for detecting ovarian cancer patients, than CEA, which gave 44% sensitivity. Also CA125 gave the highest specificity 100% for the ability of this marker to exclude normal individuals while CEA gave a specificity of 90%. The mean value of the ratio of CA125 level to CEA level in sera of ovarian cancer patients was found to be 87 in those patients with elevated CA125 concentration.

The binding of CA125 to $^{125}$I-anti CA125 antibody in the tissue of benign and malignant ovarian tumors was preliminarily tested. The results showed that the supernatant fraction of the tissue homogenates contains higher CA125 level than the pellet fraction in all studied groups.

The effects of protein concentration, $^{125}$I-anti CA125 antibody concentration, pH of the reaction medium, time of reaction and temperature were studied for the binding of CA125 to $^{125}$I-anti CA125 antibody, in the tissue homogenates of malignant and benign ovarian tumors. The optimum conditions observed for the binding were as follows:

Optimum protein concentration in tissue homogenate was (225, 150 and 175 $\mu.ml^{-1}$) for (OI, OII and OIII) respectively.

$^{125}$I-anti CA125 antibody optimum concentration was (450,360 and 450 $\mu.ml^{-1}$) for (OI, OII and OIII) respectively. Optimum pH was (7.2, 6.2 and 6.4) for (OI, OII and OIII) respectively. The optimum time and temperature was 240 minute at 4°C for all studied groups.

The use of different halides was shown to cause promotion effects on the binding of $^{125}$I-anti CA125 antibody to CA125 in groups OI, OII and OIII except I$^-$ in group OI. The studies also show that the use of different mono and divalent cations increases the binding in all groups except NH4$^+$ in group OIII.

## 2.1 Introduction

The role of ovarian tumor markers is to enhance the clinician's ability to provide more effective management of the disease. CA125 was found to be the best available marker for epithelial ovarian cancer. This oncofetal antigen is found in the embryonic coelom epithelium and, later, in the derived fetal tissues. CA125 is not expressed, or barely expressed, on the epithelial tissues of the ovary but is found in serous adenocarcinomas of the ovary from which it is shed and can then be assayed in the blood [131].

CA125 has limited specificity. It has been reported that CA125 is elevated in approximately 1% of healthy women [27], 6-40% of women with benign masses (e.g. , uterine fibroids, endometriosis, and pancreatic pseudocyst) [121] and 29% of women with non-gynecologic cancers (e.g., pancreas, lung, stomach, colon and breast) [53,108], while generally reported specificity in screening studies should be about 99% [132]. It has been reported that it may be possible to improve the specificity of CA125 measurement by either selective screening of post – menopausal women [124], modifying the assay technique, measuring other tumor markers beside CA125 [126], persistent elevation of CA125 level over time, or combining CA125 measurement with ultrasound. [125]

Serum CA125 concentration was determined by several methods including Immunoradiometric assay (IRMA) [53], Enzymeimmunoassay (EIA), Immunoflorometric assay (IFMA) [133] and enzyme-linked immunosorbent assay (ELISA) [134]. Today in commercially available CA125 assays, two monoclonal antibodies directed against different protein determinants in the CA125 protein core are used. Such antibodies seem to be less influenced by differences in glycosylation between different individuals and conditions [118].

CA125 has been used in the management of patients with ovarian cancer. It has been evaluated for its ability to determine diagnosis, prognosis, monitor therapy, predict recurrence of disease following curative surgery and treatment of ovarian cancer using immunotherapeutic strategies [135].

The objective of the present study was to evaluate the clinical application of biochemical tumor markers CA125 and CEA in the diagnosis of ovarian cancer. Also, this study was carried out in order to develop an immunoradiometric assay technique to determine optimum condition of $^{125}$I-antiCA125 antibody binding with CA125 in ovarian tumor tissue homogenates.

## 2.2 Materials

### *2.2.1 Chemicals*

All chemicals in this study were of analar grade, the specification of these chemicals are tabulated in table (2.1)

**Table 2-1: Specification of the Used Chemicals**

| No. | Chemical | Company |
|---|---|---|
| 1. | Immunoradiometric kit for CA125 antigen level. Immunoradiometric kit for CEA antigen level. | Immunotech (France) |
| 2. | Tris buffer, Bovine serum albumin, $ZnCl_2$, $CaCl_2$, $MgCl_2$, EDTA, $NH_4CL$, urea and $NaN_3$ | Fluka (Switzerland) |
| 3. | $CuSO_4.5H_2O$, Na, K –Tartarate, NaOH, HCl, $Na_2CO_3$, NaF, NaCl, NaBr, NaI, $MnCl_2$ , polyethylene glycol 6000 (PEG 6000) and Folin Ciocalteaue reagent | BDH (U.K) |
| 4. | Blue dextran, Sepharose CL-6B | Pharmacia Fine Chemicals (Sweden) |

### *2.2.2 Instruments*

**Table (2-2) Instruments Used and Companies**

| Instruments | Company |
|---|---|
| Gamma counter type 1270 rack Gamma II | LKB |
| Double Beam spectrophotometer | Shimadzu |
| pH meter | Pyeunicam |
| Sartorius analytical balance | Germany |
| Cooling centrifuge with maximum speed 5000 rpm | Hettich |
| SM shaker | England |
| Memmert water bath, Memmert incubator | Germany |

### *2.2.3 Patients*

This is a prospective study from November 2002 to September 2004 for 58 women admitted for surgical management of ovarian masses in different gynecological centers in Baghdad and southern Iraq matched with 30 healthy individuals.

Several points were taken into consideration, related to individual used throughout this study which includes the following:

(1)         No evidence of liver disease.

(2)         Not pregnant.

(3)         Un smokers.

(4)         Not taken any type of treatment.

(5)         No oral contraceptive pill used.

(6)         The time of taking the samples out of ministration period.

Patients included in this study were divided into three groups:

Group 1 consisted of 20 post – menopausal patients with ovarian cancer, group 2 consisted of 14 pre menopausal patients with ovarian cancer and group 3 consisted of 24 patients with Benign ovarian tumor. In addition to group 4 consisted of 30 healthy individuals.

The host information of all patients and normal healthy subjects is summarized in table (2-3).

**Table (2-3): The host information of ovarian tumors patient and healthy subjects Studied.**

| Group | Patients | No. | Age | Type of tumor |
|-------|----------|-----|-----|---------------|
| OI | Post-menopausal malignant ovarian tumor | 20 | 55-72 | Serous cystadenocarcinoma |
| OII | Pre-menopausal malignant ovarian tumor | 14 | 19-45 | Serous cystadenocarcinoma |
| OIII | Benign ovarian tumor | 24 | 22-46 | Benign epithelial cyst (serous cystadenoma) |
| Control | healthy individuals | 30 | 20-50 | control |

All patients were admitted for the treatment to (Medical City, Baghdad Teaching Hospital), AL- Habbibia General Hospital, Ibn Ghaswan Hospital (Basrah) and AL-Saadun private hospital (Basrah).

All surgical operations of tumor were carried out under the supervision of surgeons Dr. Fouad Al Dahhan, Dr. Luay Edward Kury and Dr. Amal Fatoohi.

## *2.2.4 Preparation of Blood Samples*

Five milliliters of blood sample were obtained from patients by vein puncture just before surgery. Blood samples were left for 20 min at room temperature after coagulation; sera were separated by centrifugation at 1500 g for 10 min. Serum specimens were then frozen at -20°C until time of analysis.

## *2.2.5 Specimens Collection*

The tumor tissues were surgically removed from ovarian tumor patient by either unilateral salpingo oophorectomy or total abdominal hysterectomy with bilateral salpingo oophorectomy. The specimens were cut off and immediately rinsed with ice-cold isotonic saline solution. They were collected individually in plastic receptacles and stored at -20°C until homogenization.

## *2.2.6 Preparation of phosphate Buffered Saline*

Phosphate buffered saline (PBS) 0.1 M, pH 7.2 was prepared as the following:

A: Disodium basic phosphate (0.1 M); 1.419 g $Na_2HPO_4$ and 0.9 g of NaCl were dissolved in a final volume 100 ml deionized distilled water.

B: Monobasic sodium phosphate (0.1M); 1.1998 g of $NaH_2PO_4$ and 0.9 g of NaCl were dissolved in a final volume 100 ml deionized distilled water.

Phosphate buffer saline pH 7.2 was prepared by mixing a volume of solution A with appropriate amounts of solution B to obtain the required pH.

## 2.2.7 Preparation of Ovarian Tumors Tissues Homogenates

The frozen tissue was weighed, sliced finely and scalped in Petri dish standing on ice bath, and then homogenized with three fold volumes of PBS buffer pH 7.2, using manual homogenizer . The homogenate was filtered through four layers of nylon gauze in order to eliminate fibers connective tissues, and then centrifuged at 1500 g for 30 min at 4°C in order to precipitate the remaining intact cells and the intact nucleus. The supernatant fraction at this speed was separated and divided in a liquots and freezed at -20°C until use

## 2.3 Methods

### 2.3.1 Determination of protein concentration

### Solutions

1. Standard bovine serum albumin (BSA), (0.2 mg/ml) as stock solution.
2. Reagent A: Alkaline carbonate solution (2% $Na_2CO_3$ in 0.1 N NaOH).
3. Reagent B: copper sulphate- sodium potassium tartarate solution (0.5% $CuSO_4$. $5H_2O$ in 1% Na, K tartarate )
4. Reagent C: Alkaline copper solution, Mixing (50ml of reagent A with 1 ml of reagent B ), discard after one day
5. Reagent D: Folin Ciocaltean reagent (1N) was prepared by the dilution of the commercial reagent (2N) with an equal volume of distilled water on the day of use.

Total homogenate protein's content was determined by the method of Lowry [136], using bovine serum albumin (BSA) as the standard solution. The details of the method are according to the following steps:

1. A volume of 1 ml of each of standard BSA (zero, 20,40, 60, 80,100, 120, 140, 160, 180, and 200) µg/ml was pipetted in a set of test tubes the experiment was carried out in duplicate.
2. A volume of 100 µl of ovarian tumors homogenate was also pipetted in test tubes and the volumes were made up to 1 ml with distilled water.

47

3.       A volume of 5 ml of reagent C was added to all assay tubes. Then the contents were mixed by vortexing and allowed to stand for 10 min at room temperature.

4.       A volume of 0.5 ml of reagent D was added drop by drop with mixing to all assay tubes the mixture was left to stand for 30 min at room temperature.

5.       The absorbance of the developing color was read at 600 nm against the appropriate blank.

6.       The standard curve was obtained by plotting the absorbance against the corresponding concentrations of standard protein and used to determine the unknown protein concentration of the sample (ovarian tumors homogenate) Fig (2-1)

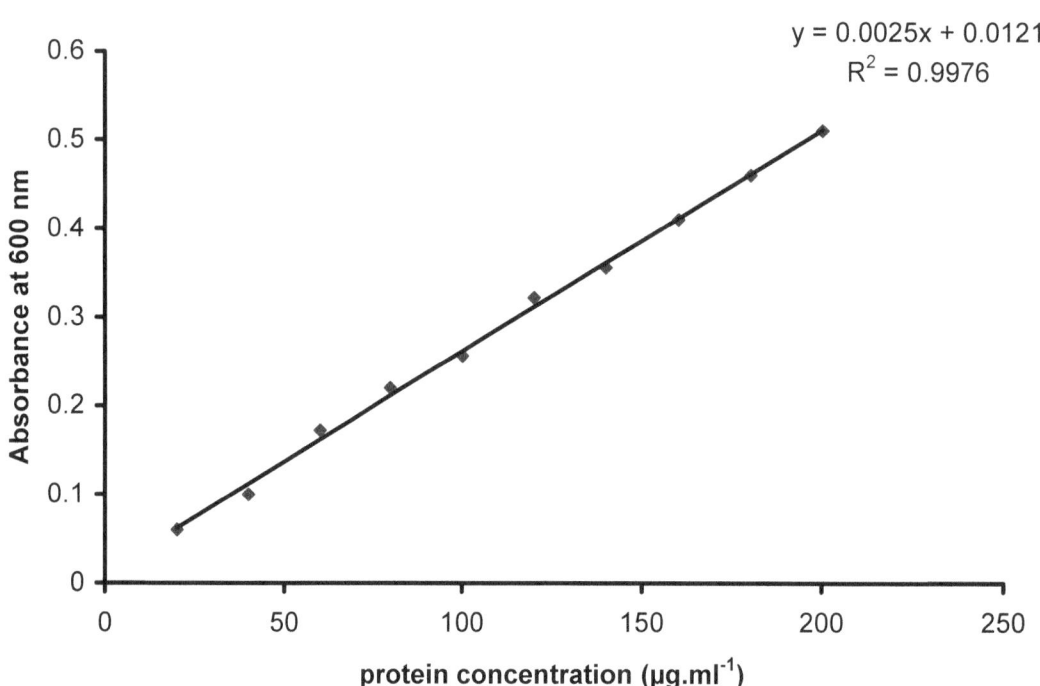

$y = 0.0025x + 0.0121$
$R^2 = 0.9976$

**Figure (2-1) standard curve of protein determination**
**(All other details are explained in the text)**

## 2.3.2 Determination of Cancer Antigen CA125 level in Sera of Patient with Benign and Malignant Ovarian Tumors

Serum CA125 levels were measured by immunoradiometric assay kit (IRMA) supplied by Immunotech (France).

IRMA – CA125 is a solid phase two-site "sandwich assay", utilizes to mouse monoclonal antibodies (OC125 and M11)directed against to different epitopes on the CA125 molecule. The M11 antibody is couted on polystyren tube (solid phase) while OC125 antibody is used as tracer after being radiolabelled with iodine 125. The CA125 antigen molecule present in the two antibodies.

Following the formation of the coated antibody/antigen/iodinated antibody sandwich, the unbound tracer is easily remove by a washing step. The radioactivity bound to the solid phase is proportional to the concentration of CA125 present in the sample.

## Reagents

The following reagents were equipped with the kit:

1. Tracer: One vial (33 ml) contains less than 480 kBq of $^{125}$I- labeled anti-CA125 in liquid form with buffer, BSA, $NaN_3$ (<0.1%).

2. Coated Tubes: Anti CA125 monoclonal antibody-coated tubes (100 plastic tubes).

3. Standard: 5 vials contain (0, 15, 50, 200 and 500 u.ml$^{-1}$ of human CA125 antigen).

4. Control serum: 1 vial (1.0 ml) of human CA125 in human serum with sodium azide (<0.1%).

5. Washing solution: 1 vial (50 ml): concentrated solution should be diluted with 950 ml distilled water before use.

## Procedure

The assay protocol is described in table (2-4).

**Table (2-4) IRMA assay protocol of CA125 u.ml$^{-1}$**

|  |  | control | unknown |
|---|---|---|---|
|  |  |  |  |

| | 0 | 15 | 50 | 200 | 500 | | 1 | 2 etc. |
|---|---|---|---|---|---|---|---|---|
| coated tubes no. | 1,2 | 3, 4 | 5,6 | 7,8 | 9,10 | 11,12 | 13,14 | 15 etc. |
| standard (µl) | 100 | 100 | 100 | 100 | 100 | | | |
| Control or sample (µl) | | | | | | 100 | 100 | 100 |
| $^{125}$I-anti-CA125 antibody* (µl) | 300 | 300 | 300 | 300 | 300 | 300 | 300 | 300 |
| | All tubes were incubated for 4 hrs at 25°C in horizontal shaker. | | | | | | | |
| | The contents of each tube were aspirated, and tubes were washed twice with 2 ml diluted washing solution except total count tubes. | | | | | | | |
| | The radioactivity bound in each tube were measured in gamma counter for 1 min. | | | | | | | |

* 300 µl of tracer were added to 2 additional tubes to obtain total c.p.m

Note: Samples having concentrations greater than the highest standard were diluted with zero standard before assay.

## Calculations

The specific binding of each concentration was measured by dividing the counts of each concentration on the total counts

$$(B/T\,\%) = \frac{\text{standard or sample mean count}}{\text{total activity mean count}} \times 100$$

The standard curve was generating by plotting the B/T% on vertical axis and the CA125 concentration of the standards on the horizontal axis (u.ml$^{-1}$)

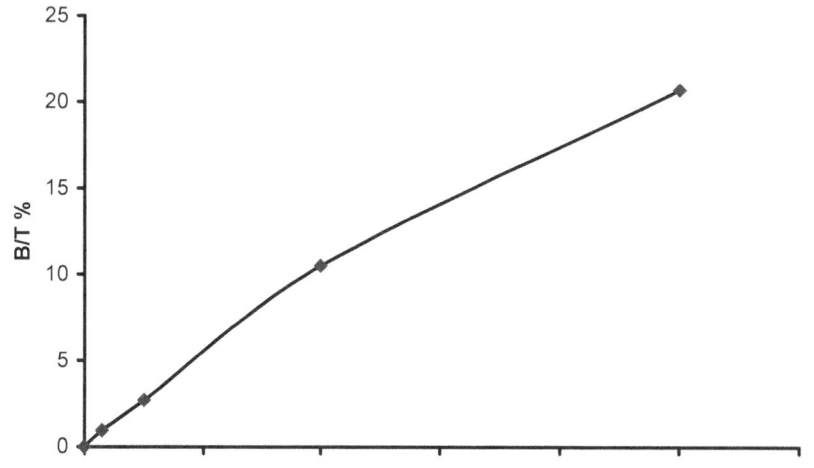

**Figure (2-2): Standard curve of CA125 determination in human sera by IRMA method (All details are explained in the text).**

### 2.3.3 Determination of Cancer Antigen CEA level in Sera of Patient With Benign and Malignant Ovarian Tumors

Serum CEA levels were measured by immunoradiometric assay kit (IRMA) supplied by Immunotech .

## Principles of the assay

IRMA – CEA is a solid phase two-site "sandwich assay ", utilizes to mouse monoclonal antibodies directed against to different epitopes on the CEA molecule. The samples or calibrators are incubated in tubes with the first monoclonal antibody in the presence of the second monoclonal antibody labeled with iodine 125.

After incubation, the content of tubes was aspirated and the tubes were rinsed so as to remove $^{125}$I- antiCA125 antibody. The bound radioactivity was then determined in gamma counter. The CEA concentration in the samples was obtained by interpolation from the standard curve and was directly proportional to the radioactivity measured.

## Reagents

The following reagents were equipped with the kit:

1. Tracer: one vial (22 ml) contains less than 640 kBq of $^{125}$I- labeled anti-CEA in liquid form with buffer, BSA, $NaN_3$ (<0.1%).

2. Coated Tubes: Anti CEA monoclonal antibody-coated tubes (100 plastic tubes).

3. Standard: 5 vials contain (0, 1, 5, 20, 100 and 400) ng.ml$^{-1}$ of human CEA antigen.

4.      Control serum: 2 vials contain CEA lyophilized in human serum.

5.      Washing solution: 1 vial contains (50 ml): concentrated solution should be diluted with 950 ml distilled water before use.

## Procedure

The assay protocol is described in table (2-5).

### Table (2-5) IRMA assay protocol of CEA ng.ml$^{-1}$

| | CEA standard in ng.ml$^{-1}$ | | | | | | Control | | Unknown samples | |
|---|---|---|---|---|---|---|---|---|---|---|
| | 0 | 1 | 5 | 20 | 100 | 400 | Level 1 | Level 2 | 1 | 2 etc. |
| coated tubes no. | 1, 2 | 3, 4 | 5, 6 | 7, 8 | 9,10 | 11, 12 | 13,14 | 15,16 | 17,18 | 19, etc. |
| standard (µl) | 50 | 50 | 50 | 50 | 50 | 50 | | | | |
| Control or sample (µl) | | | | | | | 50 | 50 | 50 | 50 |
| $^{125}$I-anti-CA125 antibody$^{*}$ (µl) | 200 | 200 | 200 | 200 | 200 | 200 | 200 | 200 | 200 | 200 |
| | All tubes were incubated for 2 hrs. at 25°C in horizontal shaker. | | | | | | | | | |
| | The contents of each tube were aspirated and tubes were washed twice with 2 ml of diluted washing solution except total counts tubes. | | | | | | | | | |
| | The radioactivity bound in each tube were measured in gamma counter for 1 min. | | | | | | | | | |

* 200 µl of $^{125}$I-antiCEA antibody were added to 2 additional tubes to obtain total count c.p.m

## Calculations

The specific binding of each concentration was measured by dividing the counts of each concentration on the total counts

$$(B/T \%) = \frac{\text{standard or sample mean count}}{\text{total activity mean count}} \times 100$$

The standard curve was generating by plotting the B/T% on vertical axis and the CEA concentration of the standards on the horizontal axis (ng.ml$^{-1}$)

**Figure (2-3): Standard curve of CEA determination in human sera by IRMA method (All details are explained in the text).**

## 2.3.4 Preliminary Test of CA125 Binding to $^{125}$I-anti CA125 antibody in Ovarian Tumor Homogenate

### Reagents

Tris buffer of 0.05 M, pH 7.4 for binding experiments was prepared according to the following:

Tris (hydroxyl methyl amino methan) 0.6075g, 0.1816g of EDTA, 1g Bovine serum albumin (BSA) and 0.02g sodium azide ( $NaN_3$ ) , were dissolved in 80 ml deionized distilled water and the pH was adjusted with HCl (1M) at pH 7.4 then the solution was completed to 100 ml with deionized distilled water.

Polyethylenglycol (PEG 6000) was prepared by dissolving 2 g in 10ml of Tris buffer (pH = 7.4, 0.05 M).

### Procedure

1- Five clean and dry tubes were counted for their background using Gamma counter.

2- Twenty five microliter (450 $\mu$g.ml$^{-1}$) of $^{125}$I-anti CA125 antibody (tracer) was added to each of five tubes denoted as

First tube : Non specific binding without precipitating agent.

Second tube: Non specific binding with precipitating agent.

Third tube : Binding without precipitating agent

Fourth tube : Binding with precipitating agent

Fifth tube : Total c.p.m (T)

3- Filtrate of homogenate, 50 $\mu$l (337 $\mu$g) was added to third and fourth tubes.

4-          The volume of all tubes was completed to 400 µl using Tris-buffer pH 7.4, 0.05 M except the T tube.

5-          The tubes were incubated at 25°C for 4 hrs.

6-          After incubation 400 µl of 20% polyethyleneglycol 6000 (PEG 6000) was added to the second and fourth tubes and the incubation was continued for further 30 min.

7-          After incubation all tubes were centrifuged at 1500 g for 30 min. at 4°C except (T) tube.

8-          The supernatant was aspirated, except (T) tube

9-          The radioactivity in each tube was counted using gamma counter for 1 min.

10-        The pellet of the homogenate was suspended in 1:3 Tris-buffer pH 7.4, 0.05 M, and the same steps mentioned above were repeated for the suspended pellet to determine the radioactivity of the complex.

## Calculations

1.     The counts radioactivity in tubes number 2, 1 (expressed in c.p.m.) represents the non specific binding (NSB) with and without using precipitating agent respectively.

2.     The counts radioactivity in tubes number 4, 3 (expressed in c.p.m.) represents the sample count with and without using precipitating agent respectively.

3.     (Sample counts-non specific binding counts) represents bound fraction (B).

4.     The counts radioactivity in the tubes containing [125]I-anti CA125 antibody only represents the total counts (T). The (B/T) ratio counted as follows:

$$\frac{B}{T}\% = \frac{\text{Sample counts} - \text{non specific binding counts (NSB)}}{\text{Total counts (T)}} \times 100$$

## 2.3.5 Factors Affecting the Binding CA125 to $^{125}$I-anti CA125 antibody in ovarian tumor Homogenate.

### 2.3.5.1 Effect of Different Amount of protein concentration of the tumor Homogenate on the Binding of CA125 with $^{125}$I-anti CA125 antibody.

### Reagents:

All reagents were prepared as described previously in section (2.3.4).

### Procedure

1.      Twenty five microliters (450 µg.ml$^{-1}$) of $^{125}$I-anti CA125 antibody were read for their radioactivity and added to an increasing amounts of protein (37.5, 75, 150, 175, 225 and 300 µg.ml$^{-1}$) of the supernatant of (post-menopausal malignant ovarian tumors "OI", pre-menopausal malignant ovarian tumors "OII", and benign ovarian tumors "OIII").
Then all volums were completed to 400 µl with Tris buffer (0.05M, pH 7.4).

2.      The assay tubes were then incubated for 4 hrs at 25°C.

3.      At the end of the incubation 400 µl of 20% polyethyleneglycol 6000 (PEG 6000) was added and the incubation was continued for further 0.5hr.

4.      At the end of incubation, the assay tubes were centrifuged at 1500 g for 30 min at 4°C.

5.      Supernatant was aspirated carefully.

6.      The radioactivity of the complex was counted using gamma counter.

### Calculations

1.      The B/T % was calculated as described in section (2.3.4).

2.      The B/T% was plotted against the increasing amount of protein of ovarian tumor homogenate

## 2.3.5.2 Effect of $^{125}$I-anti CA125 antibody concentration on its binding to CA125

### Reagents

All reagents were prepared as described previously in section (2.3.4).

### Procedure

1.	Increasing concentrations of $^{125}$I-anti CA125 antibody (90, 180, 270, 360, 450, and 900 µg.ml$^{-1}$) were read for their radioactivity and then added to homogenate (OI, OII, OIII) containing (225, 150, 175 µg.ml$^{-1}$) respectively. Then all tubes were completed to 400 µl with Tris-buffer (0.05M, pH 7.4).

2.	Steps 2, 3, 4, 5 and 6 mentioned in the experiment (2.3.5.1) were repeated.

### Calculations

1.	The B/T% was determined using the same mathematical equation mentioned in section (2.3.4).

2.	The percent of binding was plotted against $^{125}$I-anti CA125 antibody concentrations.

## 2.3.5.3 The Effect of pH on Binding of CA125 to $^{125}$I- anti CA125 antibody

### Reagents

All reagents were prepared as mentioned in section (2.3.4), except PEG 6000. PEG 6000 which was prepared in a set of different pHs (6.0-8.0) by dissolving 2 g of PEG 6000 in 10 ml of Tris-buffer of different pHs (6.0-8.0).

### Procedure

1. Tracer (450, 360, 450 µg.ml$^{-1}$) were added to (225, 150, 175 µg.ml$^{-1}$) of OI, OII and OIII homogenate respectively.

2. Each reaction mixture was completed to 400 µl with Tris-buffer at different pH (6.0-8.0).

3. Steps 2, 3, 4, 5 and 6 of the experiment 2.3.5.1 were repeated.

## Calculations

1. Values of B/T % were calculated as described in section (2.3.4)

2. B/T % was plotted against their pH values.

### 2.3.5.4 Time course of the binding of CA125 with $^{125}$I-antiCA125 antibody in ovarian tumor homogenate

## Reagent

All reagents were prepared according to the experiment in section (2.3.4) except 20% PEG 6000 solution which was prepared according to the optimum pH of each group (i.e. OI, OII and OIII).

## Procedure

1. Tracer (450, 360, 450 µg.ml$^{-1}$) were added to (225, 150, 175 µg.ml$^{-1}$) of OI, OII and OIII homogenate respectively.

2. Each mixture was completed to 400 µl with Tris -buffer at the optimum pH of each group

3. All tubes were incubated at 25°C at different time intervals (0.5, 1, 1.5, 2, 2.5, 3, 3.5, 4, 4.5, and 5) hrs.

4. Steps 3, 4, 5 and 6 in the experiments (2.3.5.1) were repeated.

5. To determine the time course of CA125 binding to $^{125}$I-anti CA125 antibody at different temperatures.

Steps 1, 2, 3, 4 and 5 in this experiment were repeated at different temperature (5, 37 and 45) °C.

## Calculations

1.       The same mathematical equation mentioned in section (2.3.4) was used to calculate (B/T) % at each time and temperature.

2.       The (B/T) % values were plotted against the time of incubation at different temperatures.

### 2.3.5.5 Effect of Different Halides on the Binding of CA125 to $^{125}I$-anti CA125 antibody.

### Reagents

1.       Tris buffer, that was prepared as described in section (2.3.4), was adjusted to corresponding pH for each group of tissue homogenate.

2.       Halides solution was prepared in concentration of (0.01M) in Tris buffer at pH (7.2, 6.2 and 6.4) individually, by dissolving each of 0.021 gm of NaF, 0.0292 gm of NaCl , 0.0515 gm of NaBr, and 0.075 gm of NaI in final volume of 50 ml of Tris buffer and the pH was adjusted.

3.       The ovarian tumor homogenates were prepared as described in section (2.2.7), except phosphate buffer was used instead of phosphate buffer saline at the same pH and concentration was used as homogenizer buffer.

### Procedure

1.       Tracer (450, 360, 450 $\mu g.ml^{-1}$) were added to (225, 150, 175 $\mu g.ml^{-1}$) of protein of (OI, OII and OIII) homogenate respectively.

2.          Fifty micro liters of the following halides (0.01 M) (NaI, NaBr, NaCl and NaF) were added in each assay tube. (A sample without the addition of any salt was used as a control).

3.          The volume of the mixture was completed to 400 µl with Tris-buffer at the optimum pH of each group.

4.          The assay tubes were then incubated for 240 minute at 5°C for all studied groups.

5.          Steps 3, 4, 5 and 6 mentioned in section (2.3.5.1) were repeated.

## Calculations

1.          The values of (B/T) % were calculated as described in section (2.3.4).

2.          (B/T) % was plotted against the halide type.

### 2.3.5.6 *Effect of Monovalent and Divalent Cation on the binding*

## Reagents

1.          Tris buffer was prepared as described in section (2.3.4) was adjusted to corresponding pH for each group of tissue homogenate.

2.          Monovalent and divalent cations salts were prepared in concentration of (0.025M) in Tris buffer at pH (7.2, 6.2 and 6.4) individually, by dissolving each of 0.0931 gm of KCl, 0.0668 gm of $NH_4Cl$, 0.2541 gm of $MgCl_2.6H_2O$, 0.1388 gm of $CaCl_2.2H_2O$, 0.2474 gm of $MnCl_2$ $4H_2O$, 0.315 gm of $CuSO_4$. $5H_2O$ and 0.1703 gm of $ZnCl_2$) in a final volume 50 ml of Tris and the pH was adjusted.

## Procedure

1.          Step (1) of effect of halide experiment was repeated.

2.          Fifty micro liters of (0.25 M) of the following monovalent and divalent cations (KCl, $NH_4Cl$, $MgCl_2$ $6H_2O$, $CaCl_2.2H_2O$, $MnCl_2.4H_2O$, $CuSO_4.5$ $H_2O$ and $ZnCl_2$) were added to each group of tissue homogenate.

3.          Steps 3, 4, 5, and 6 in effect of halide experiment were repeated.

## Calculations

1. The values of (B/T) % were calculated as described in section (2.3.4).

2. B/T % was plotted against each monovalent and divalent cations.

## 2.4. Result and Discussion

Three groups of ovarian tumors were included in this study. These groups were classified according to the type of ovarian tumors (benign and malignant) and the malignant ovarian tumors were again classified into sub groups (pre-menopausal and post-menopausal). Each type was examined histologically according to WHO classification system.

Homogenization of tissue samples was carried out in cold medium ($4^0$C) to avoid protein denaturation and to decrease the proteolytic enzymes activity. The filtration of the tissue homogenate through several layers of nylon gauze was used to remove any suspended piece of unhomogenized fragments and blood vessels, while the centrifugation of homogenate at 1500 g removed the unruptured cells and intact nuclei of ruptured cells .

## 2.3.2 Determination of Cancer Antigen CA125 level in Sera of Patient with Benign and Malignant Ovarian Tumors

CA125 levels in sera were measured with an Immunoradiometeric assay (IRMA) in three groups of ovarian tumors matched with one group of control subjects. Group I consisted of twenty post-menopausal patients with malignant ovarian tumors, group II consisted of fourteen pre-menopausal patients with malignant ovarian tumors and group III consisted of twenty four pre-menopausal patients with benign ovarian tumors.

The data of CA125 measurements in normal healthy individuals, benign ovarian tumors and malignant ovarian tumors will be presented separately.

### Normal controls

Low levels of CA125 were observed in the sera of 30 apparently healthy women used as a control (Table 2-4) . The mean CA125 levels ($\pm$SD) in this group was ($11.9 \pm 6.6$ u.ml$^{-1}$) with an upper normal value of   35 u.ml$^{-1}$ .

**Table (2-6): Sera CA125 levels (u.ml$^{-1}$) in patients with benign and**

| Group | Patients | No. | Age range | CA125 assay u.ml⁻¹ | | | P values |
|-------|----------|-----|-----------|-------|--------|---------|----------|
| | | | | Range | Median | mean ± SD | |
| OI | Post-menopausal malignant ovarian tumor | 20 | 55-72 | 12-2300 | 288 | 512± 621 | P<0.05 |
| OII | Pre-menopausal malignant ovarian tumor | 14 | 19-45 | 10-610 | 161 | 212± 200 | P<0.05 |
| OIII | Benign ovarian tumor | 24 | 22-46 | 5-50 | 13 | 19.7± 13.93 | P<0.05 |
| Control | Healthy individuals | 30 | 20-50 | 3-28 | 11 | 12.9± 6.6 | |

P – value ≤ 0.05 is consider significant.

A positive scoring or an abnormal level was indicated by those values of CA125 which exceeded the 35u.ml⁻¹ limited [53]. All normal controls had CA125 concentration lower than 35u.ml⁻¹ suggesting a test specificity of 100% for the ability of this marker to exclude normal individuals.

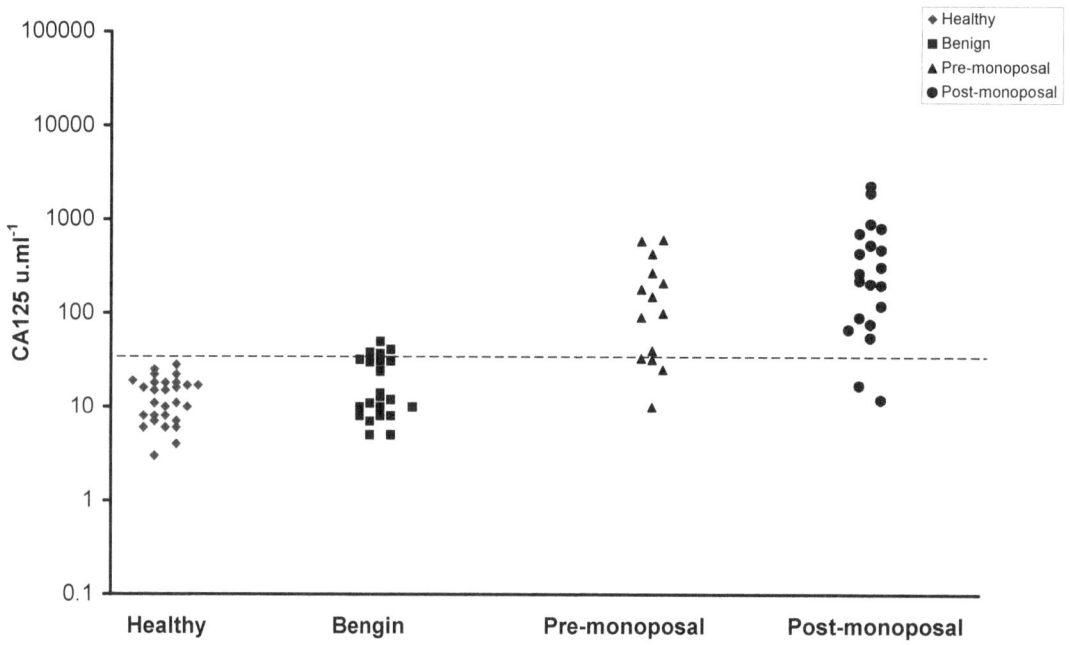

**Figure (2-4): Distribution of CA125 value over different groups of patients and healthy individuals.**
**(All other details are explained in the text).**

## Benign ovarian tumor

Figure (2.4) shows that of 24 samples of patients with histological confirmed benign ovarian cases, four had level of CA125 above the cutoff value >35u.ml$^{-1}$ (about 16%). The mean level of CA125 antigen (19.7±13.93) observed in these patients (Table 2-6 ) was significantly different from normal controls (P<0.05). Fleurn et al [137] explained why most benign cases have CA125 level below 35u.ml$^{-1}$ as that in benign ovarian tumor there might be an effective barrier between the antigen-producing neoplastic cells and the serum. Such barriers may play a role in the distribution of tumor antigens over various body compartments whereas in malignant tumors infiltrative growth lead to the release of antigen into the circulation [137].

Another explanation suggests that peritoneum serves as a barrier for high molecular weight tumor antigens, and lag in the transfer of CA125 from Cyst fluid through the cyst wall might be increased by the high molecular weight of this glycoprotein which has been estimated to exceed 200 kDa [137].

Slight elevation of CA125 serum levels in few patients with benign ovarian tumors may be explained by the observations of Bast et al which assumed that high antigen levels in cyst fluid might be able to establish a concentration gradient favouring diffusion of antigen into the lymph vessels and veins of the ovary[117].

## Ovarian Cancer

The data in table (2-6) show that the mean serum value ±SD of CA125 in pre-menopausal malignant ovarian tumor patients (212 ± 200u.ml$^{-1}$), was significantly higher than that in healthy control or patients with benign ovarian tumor. These results are in agreement with that found by AL-Barazanji [138] who obtained the same results using Mini VIDAS CA125 II kit.

Result in table (2-6) also indicates that there is significant difference between mean serum value ±SD of CA125 in post- menopausal ovarian cancer patients ($512 \pm 621$ u.ml$^{-1}$) and that found in pre- menopausal ovarian cancer patients (P<0.05) that is the same finding of Malkasione et al study [139].

Data in table (2-6) indicate that determination of CA125 serum levels may be useful as a prognostic factor in the differentiation between malignant and benign ovarian cancer especially in postmenopausal patients. These results were in agreement with the observations reported by other investigators which suggested that among postmenopausal patients with pelvic mass, CA125 level greater than 65u.ml$^{-1}$ indicated the presence of malignant disease with greater than 90% accuracy [134].

In table (2-7) the percentage of positive scoring of CA125 is presented in relation to menopausal status of ovarian cancer women. The post menopausal patients gave the highest percentage 90% of CA125 positive scoring, in comparison to the pre- menopausal patients with 71% sensitivity.

**Table (2-7): positively of CA125 in relation to menopausal status**

| Menopausal state | No. of patients | No. of elevated CA125 level | % Patients with elevated CA125 |
|---|---|---|---|
| Pre- menopausal | 14 | 10 | 71 |
| Post- menopausal | 20 | 18 | 90 |

## 2.4.2 Determination of CEA level in sera of ovarian tumor patients

Tumor markers complementary to CA125 were mentioned as a useful method to improve the specificity of CA125 assay. A specificity of 99.6% and sensitivity of 80% were reported to be achieved using OVXI antigen as a complementary tumor marker to CA125 in the diagnosis of ovarian cancer. [126] CEA levels have been reported to be elevated in a high percentage of ovarian cancer cases (30-65%).[53] High ratio of CA125 to CEA in serum was suggested to be a useful method to differentiate ovarian from non-ovarian malignant diseases when both sera contain increased CA125 concentrations. [140]

## Normal Controls

Low levels of serum CEA were observed in normal women (n=30) who had a mean value (±SD) $1.2 \pm 0.82$ ng.ml$^{-1}$ with the cut off value of 3.0 ng.ml$^{-1}$ (Table 2-8), and a percentage specificity of 90%. These values are close to those obtained by other investigators. [141, 142]

**Table (2-8): Sera CEA Levels (ng.ml$^{-1}$) in Patients with Benign and Malignant Ovarian Tumors Compared to the Control Group.**

| Group | Patients | No. | Age range | CEA assay ng.ml$^{-1}$ | | | P values |
|---|---|---|---|---|---|---|---|
| | | | | Range | Median | mean ± SD | |
| OI | Post-menopausal malignant ovarian tumor | 20 | 55-72 | 0.9-15 | 2.8 | 4.1± 3.66 | P<0.05 |
| OII | Pre-menopausal malignant ovarian tumor | 14 | 19-45 | 0.8-10 | 2.85 | 3.39± 2.35 | P<0.05 |
| OIII | Benign ovarian tumor | 24 | 22-46 | 0.5-5.5 | 1.45 | 1.6 ± 1.37 | P>0.05 |
| Control | Healthy individuals | 30 | 20-50 | 0.3-3.6 | 0.9 | 1.2± 0.82 | |

P – value $\leq 0.05$ is consider significant.

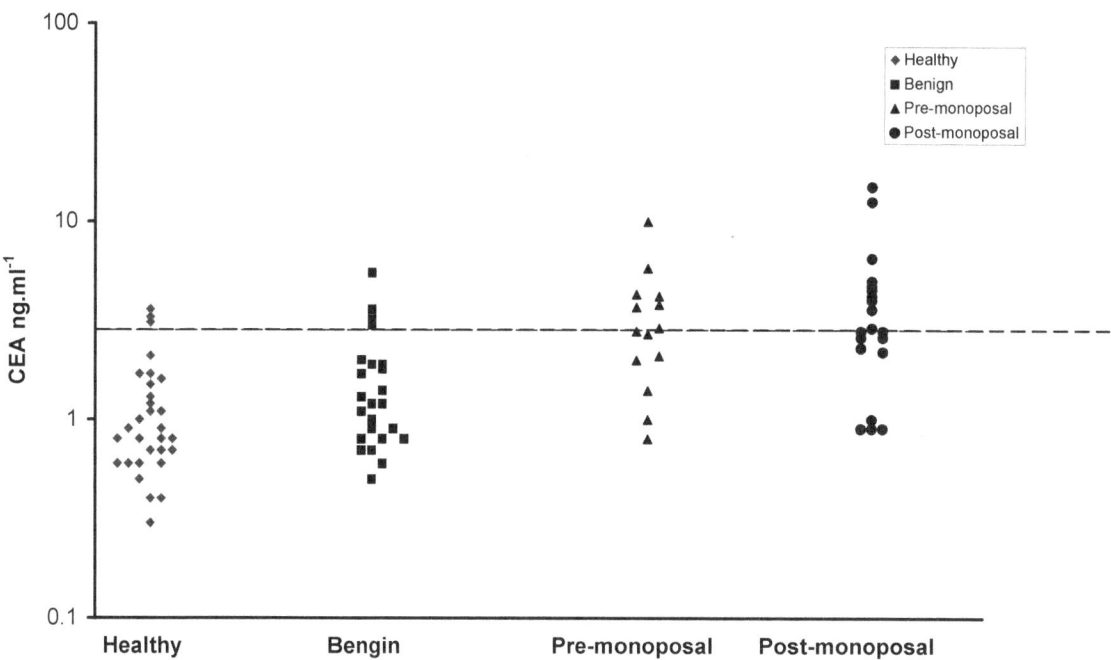

**Figure (2-5): Distribution of CEA value over different groups of patients and healthy individuals.**
**(All other details are explained in the text).**

## Benign Ovarian Tumor

Of 24 samples from patients with histologically confirmed benign ovarian cases(Fig 2-5), four of them had CEA >3.0 ng.ml$^{-1}$. The other 20 patients gave CEA value <3.0 ng.ml$^{-1}$, which is not significantly different from normal controls (P>0.05).

## Ovarian Cancer

The data in table (2-8) show that the mean serum value (±SD) of CEA in pre-menopausal patients with ovarian cancer is (3.39±2.35 ng.ml$^{-1}$) which is significantly higher (P<0.05) than their correspondent values in sera of healthy and that in patient with benign tumors, while there were insignificant differences (P>0.05) between CEA values in patients with benign tumor in comparison to their values in healthy control.

Table (2.9) shows that the rate of positive scoring was found not remarkably affected by menopausal states. It was 45% in post-menopausal states in comparison to 43% in pre-menopausal states.

The sensitivity of CEA tumor marker was found 44%, which is lower than that found by Donaldson et al. (60%) that may be due to different cell types included in each study. [143]

**Table (2-9): positively of CEA in relation to menopausal status**

| Menopausal state | No. of patients | No. of elevated CEA level | % Patients with elevated CEA |
|---|---|---|---|
| Pre- menopausal | 14 | 6 | 43 |
| Post- menopausal | 20 | 9 | 45 |

## *2.4.3 Determination of the ratio of CA125 level to CEA level in Sera of ovarian cancer patients*

Of 34 samples from patients with histologically confirmed malignant ovarian cases, 29 samples contain increased CA125 concentration (85%). High ratio of CA125 level to CEA level in sera of patients with increased CA125 concentration were found 87, in comparison to 0.94 for non-ovarian malignances (colorectal, breast, lung, and pancreatic carcinomas), reported by James et al. [140]

This observation confirms the results found by James et al. which suggest that high ratio could be used to differentiate ovarian from non ovarian malignant diseases when both sera contain increased CA125 concentration, and that will really improve the specificity of the CA125 test for ovarian cancer.

## 2.4.4 Preliminary test of the binding of CA125 to $^{125}$I-anti CA125 antibody in Ovarian tumor homogenate.

This part of the work was carried out to check the assay method in order to be able to find out the optimum conditions for CA125 binding to its specific antibody in our studied women. Supernatant and pellet formed at speed 1500 g in the three groups of human ovarian tumor homogenate (benign ovarian tumor, pre and post-menopausal malignant ovarian tumors) were used in this experiment. Table (2-10) shows the results of the preliminary test for the binding.

**Table (2-10):**  **Incidence of CA125 in Supernatant and Pellet fractions in the three fferent Ovarian homogenate**

| Group | B/T % | | | | | | | |
|---|---|---|---|---|---|---|---|---|
| | supernatant | | | | Pellet | | | |
| **Post-menopausal (OI)** | $NSB^-$ | $(NSB)^+$ | $B^-$ | $B^+$ | $NSB^-$ | $(NSB)^+$ | $B^-$ | $B^+$ |
| | 1.2 | 1.21 | 2.1 | 12.2 | 1.25 | 1.19 | 1.9 | 4 |
| **Pre-menopausal (OII)** | 1.07 | 1.1 | 1.1 | 5.3 | 1.1 | 1.1 | 1.13 | 1.6 |
| **Benign(OIII)** | 1.17 | 1.2 | 1.7 | 3 | 1.7 | 1.18 | 1.15 | 1.1 |

NSB$^-$      : nonspecific binding in absence of precipitating agent.
(NSB) $^+$  : non specific binding in presence of precipitating agent.
B$^-$        : binding in absence of precipitating agent.
B$^+$        : binding in presence of precipitating agent.

The results reveal that the supernatant fraction contains higher CA125 content than the pellet fraction according to the (B/T%) values, therefore the pellet fraction was discarded. Complex ($^{125}$I-antiCA125 antibody/CA125) formed did not precipitate in the absence of the precipitating agent . So it is necessary to use (20% PEG 6000) in the reaction mixture to precipitate this complex .

## 2.4.5 Factors Affect of $^{125}$I-anti CA125 antibody binding to CA125 in ovarian tumor homogenates

### 2.4.5.1 The Effect of different amounts of protein concentration of the tumor homogenate on the binding with $^{125}$I- anti CA125 antibody.

To obtain the optimum concentration of homogenates for the binding of CA125 with $^{125}$I-anti CA125 antibody, the supernatant of the homogenate containing increasing amounts of CA125 were incubated with a fixed amount of $^{125}$I-anti CA125 antibody, according to the details in section (2.3.5.1).

Figure (2-6) represents the quantities precipitation curve in which the amount of ($^{125}$I-antiCA125 antibody/CA125) complex in three groups (benign ovarian tumors, pre-and post-menopausal ovarian tumors) was plotted as a function of CA125 concentration.

The results revealed that the binding of CA125 to $^{125}$I-antiCA125 antibody increases with increasing CA125 homogenate until a point of maximum binding was reached, thus the increase in protein concentration which would increase the number of binding sites until the saturation state. After this point as the amount of CA125 increased the amount of complex formed diminished that means the reaction behaves according to Hook effect which has ascending and descending phases at low and high antigen concentration [144]. The decrease in binding after reaching the maximum binding may be due to the conformational changes in CA125 and $^{125}$I-antiCA125 antibody [145].

According to the results obtained in this experiment the amount of (225, 150, 175 μg.ml$^{-1}$) of tissue homogenate in groups OI, OII and OIII respectively were used in all subsequent experiments.

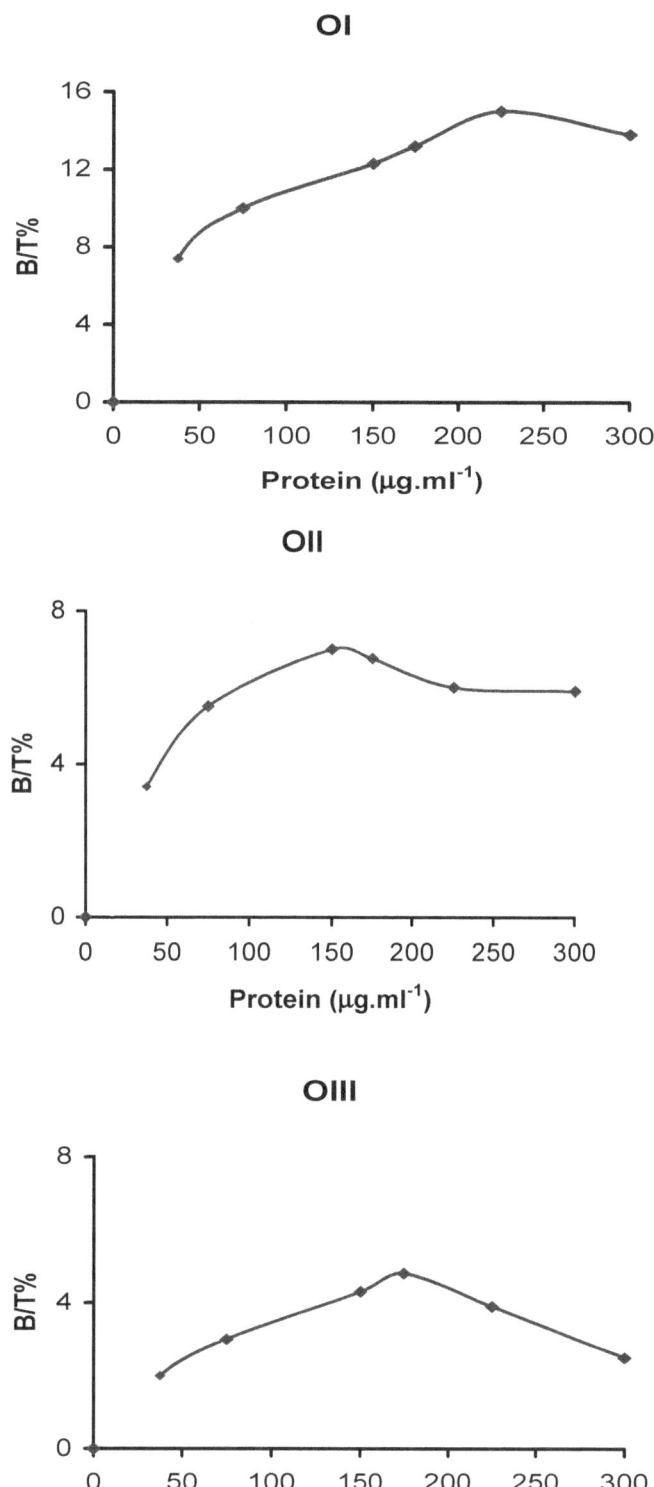

**Figure (2-6): The effect of protein concentration on the binding of CA125 to its $^{125}$I-antiCA125 antibody in ovarian tumors homogenates OI, OII and OIII (All other details are explained in the text).**

## 2.4.5.2 Effect Of $^{125}$I-antiCA125 antibody concentration on the binding to CA125.

The experiment was carried out in the presence of fixed amount of protein concentration of the homogenate and increasing concentration of $^{125}$I-antiCA125 antibody.

The results are illustrated in figure (2-7) which represents $^{125}$I-antiCA125 antibody binding with supernatant fraction of the three studied groups. As shown in figure (2-7) it is obvious that the amount of ($^{125}$I-antiCA125 antibody/CA125) complex rises gradually, and then the ovarian tumor protein was saturated with $^{125}$I-antiCA125 antibody. When the amount of antibody is in moderate excess, the probability of cross-linking of $^{125}$I-antiCA125 antibody to CA125 in the incubation mixture is more likely, and hence large complex formation is favoured then the maximum B/T percent was detected. After that the binding percent decreased as the amount of $^{125}$I-antiCA125 antibody increased, the reason is that all antigenic sites covered with antibody and complex formation is inhibited [146].

According to the results obtained in this experiment the amount of (450, 360, 450, µg.ml$^{-1}$) of $^{125}$I-antiCA125 antibody in the three studied groups (OI, OII and OIII) respectively were used in all subsequent experiments.

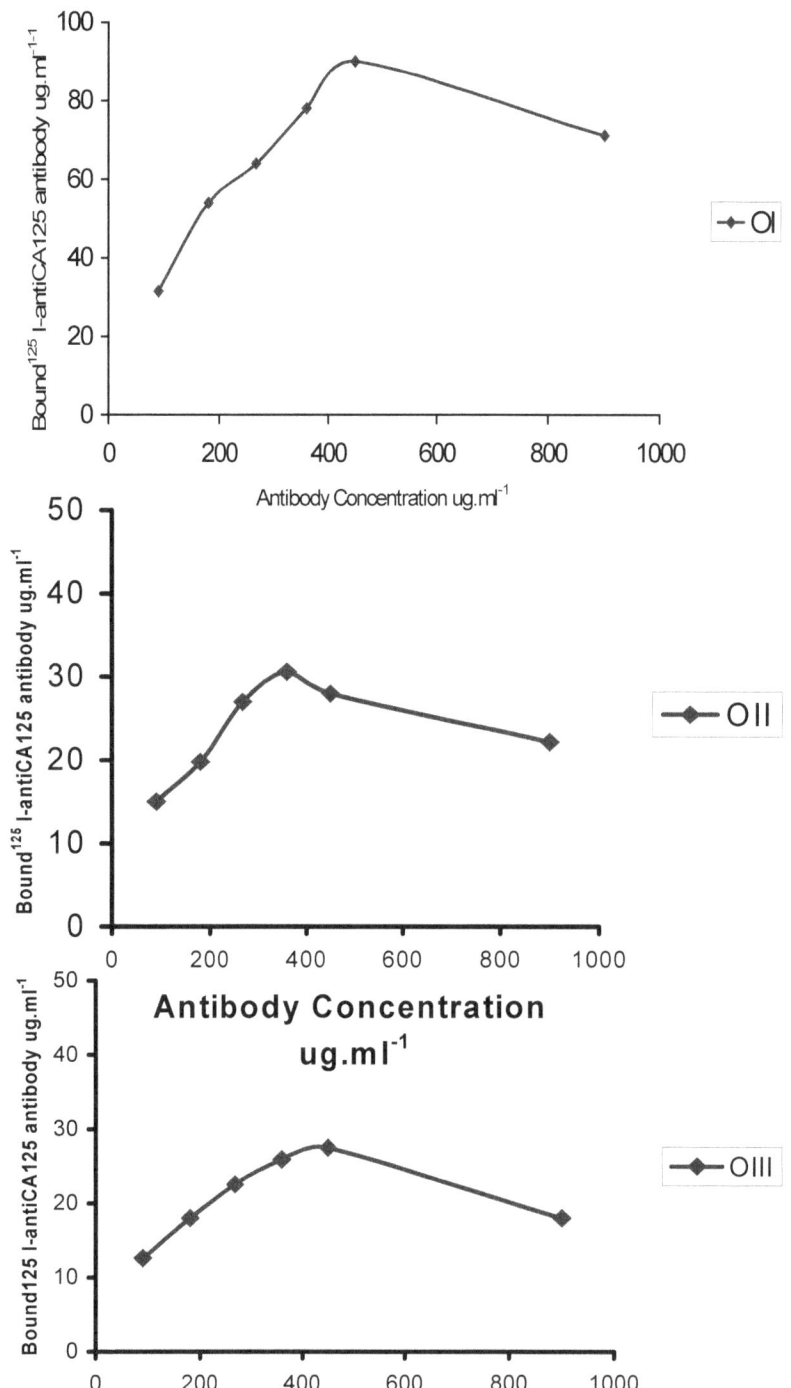

**Figure (2-7): The effect of** [125]**I-antiCA125 antibody concentration on the binding with CA125 Antigen in ovarian tumors homogenates OI, OII and OIII (All other details are explained in the text).**

## 2.4.5.3 *Effect of* pH *on binding of CA125 with* [125]*I-anti CA125 antibody*

Figure (2-8) shows the value of the binding of CA125 to [125]I-anti CA125 antibody in the three studied groups (OI, OII and OIII) at different pH values.

Maximum value of the binding occurs at pH 7.2 for the binding of post-menopausal malignant ovarian tumors with [125]I-anti CA125 antibody and pH 6.2, pH 6.4 for OII and OIII groups respectively, these results are in agreement with many protocols developed to detect CA125 antigen in serum samples using immunoradiometric assay at optimum reaction pH around 6.0 [53, 97, 133]. These results indicate that the binding was pH dependent and the shift in the pH of the environment may affect the properties of CA125 molecules involved in the binding, this effect may include protonation deprotonation processes occurring within the possible ionizable groups of the amino acids present in the binding domain of these molecules [147]. In addition, [125]I-antiCA125 antibody itself may have ionizable groups and only at a certain pH the antibody will have ionic form where it can bind to CA125.[148]

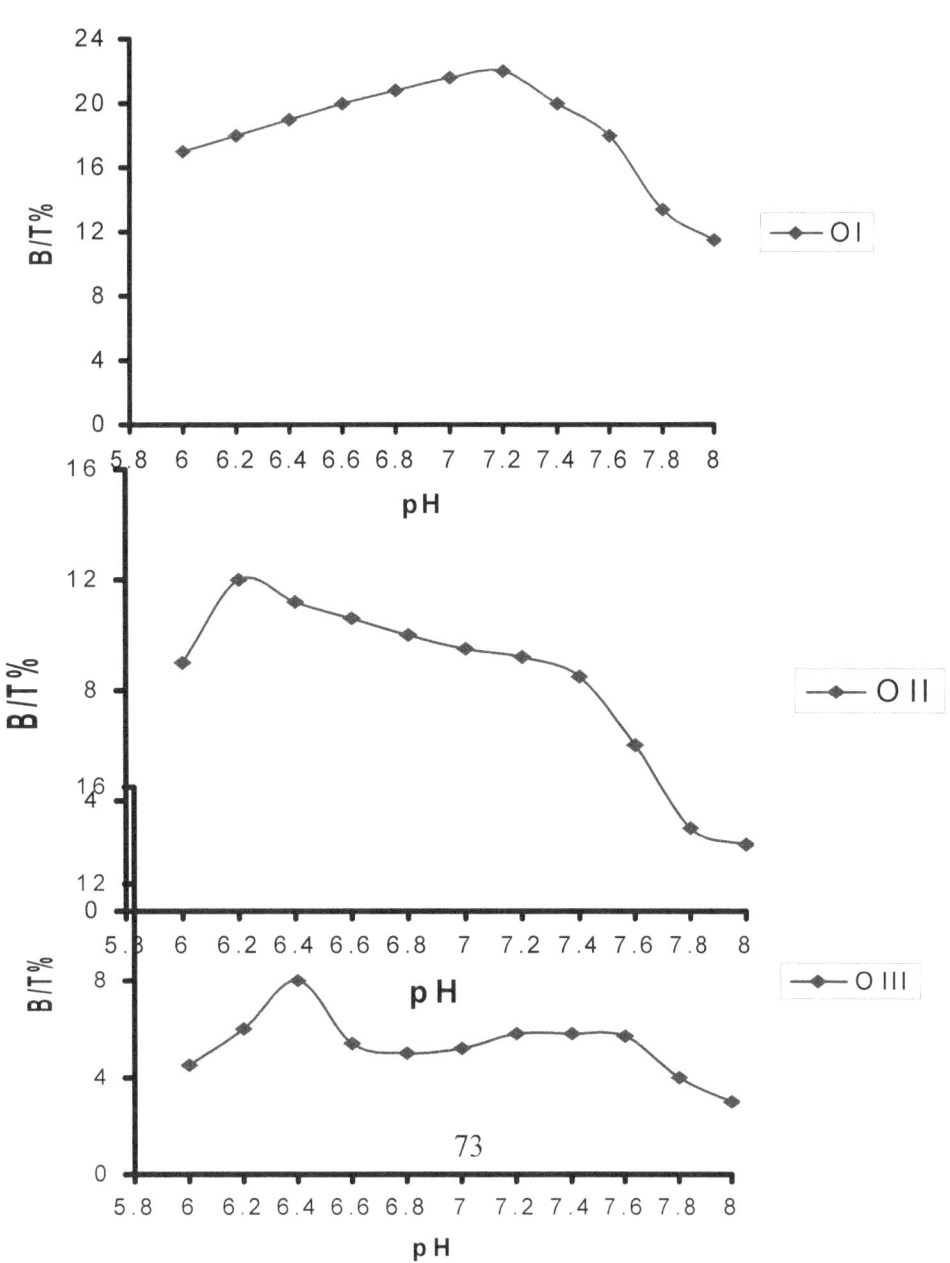

**Figure (2-8): The effect of pH on the binding with CA125 to its $^{125}$I-anti CA125 antibody in ovarian tumors homogenates OI, OII and OIII (All other details are explained in the text).**

### 2.4.5.4 Time course of the binding of CA125 to $^{125}$I-antiCA125 antibody in ovarian tumor homogenate

Figure (2-9) shows the time course of binding of CA125 to $^{125}$I-antiCA125 antibody at different temperatures (5, 20, 37 and 45°C). The maximum binding occurred at 5°C after incubation for 4 hrs in crude fractions of post and pre menopausal malignant ovarian tumor and at 5°C after incubation for 4.5 hrs for benign ovarian tumor homogenate, in these states it seems that the energy is enough to overcome the energy barrier and give maximum binding[149].

The results indicate that $^{125}$I-antiCA125 antibody binding to crude fraction of CA125 is temperature and time dependent process.

The decrease in the binding as temperature increase may be due to either degradation of CA125 or irreversible dissociation of the ($^{125}$I-antiCA125 antibody/CA125) complex at higher temperature, denaturation and destruction tertiary structure may occur leading to conformational changes and loss of activity.

The results obtained in these experiments were used in all subsequent experiments.

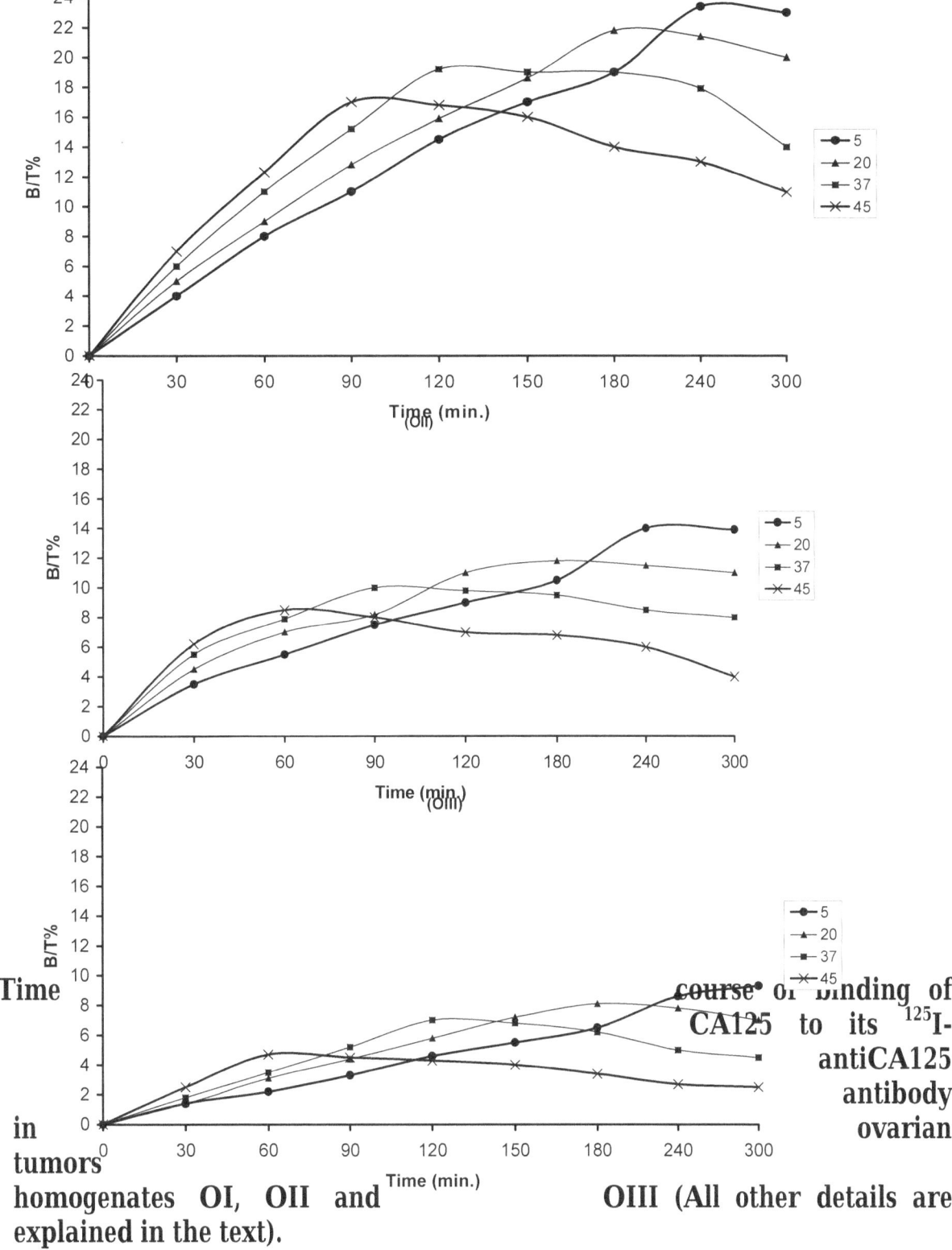

**(9):** **Time** course of binding of CA125 to its [125]I-antiCA125 antibody in tumors homogenates OI, OII and OIII ovarian (All other details are explained in the text).

### 2.4.5.5 The Effect of Different Halides on the Binding

Figure (2-10) shows the effect of different sodium halides (i.e NaF, NaCl, NaBr, and NaI) at 0.01 M concentration on the binding of [125]I-antiCA125 antibody with CA125 in benign ovarian tumors and pre and post-menopausal malignant ovarian tumors.

It seemed that the sodium halides promoted the binding according to the following order:

NaI < NaBr < NaCl < NaF

The order corresponds to the decreasing ionic radius and increasing radius of hydration presumably, the lesser degree of hydration permits greater interaction of the salt with an ionic group located in the antigen or antibody. [11]

Melander and Horvath [150] reported that the capacity of the halides salt was due to the influence of hydrophobic interaction and dependence on the molal surface tension increment (MSTI), the halides with higher MSTI strengthens the hydrophobic interaction, while halide with lower MTSI values reverses this effect.

The magnitude of surface-tension increment depends on the interaction of salt ions with the surrounding water. High – lytropic series salts (kosmotropes) interact with water strongly; water molecules surrounding the salt ions are more structured relative to bulk water. Low-lyotropic series salts (chaotropes) break the structure of the surrounding water molecules (relative to the bulk water) as a result of the large size of the ion and its weak interaction with water. [151]

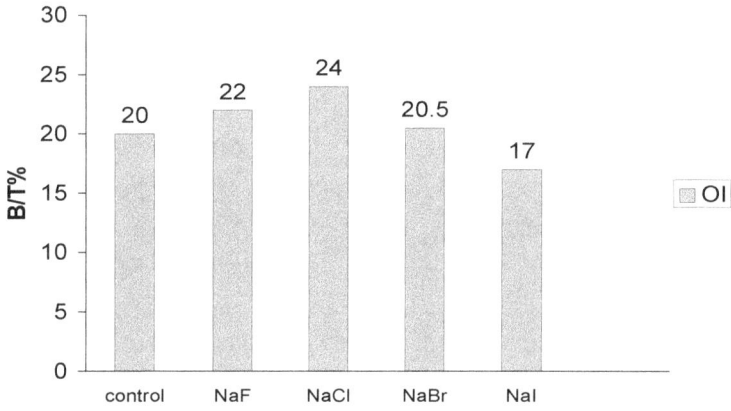

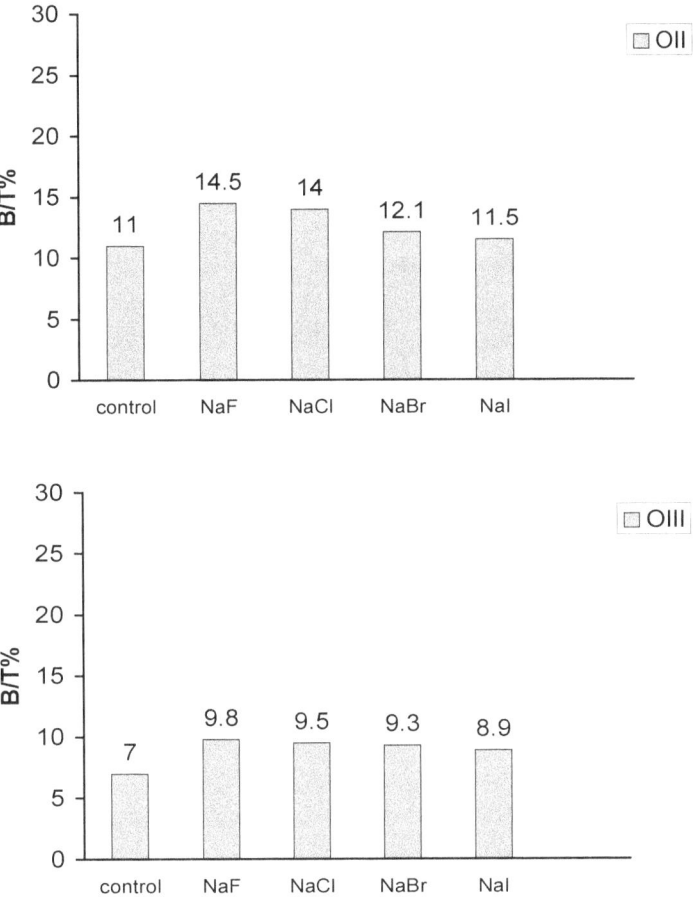

0): **The effect of different halides on the binding of CA125 to its $^{125}$I-antiCA125 antibody in ovarian tumors homogenates OI, OII and OIII (All other details are explained in the text).**

### 2.4.5.6 The Effect of Monovalent and Divalent Cations on the binding.

The effect of different salts on the extent of binding of $^{125}$I-antiCA125 antibody to CA125 in benign ovarian tumors, post and pre-menopausal malignant ovarian tumors are shown in figures (2-11) and (2-12).

The result indicates that the presence of divalent cations (i.e. $MgCl_2.6H_2O$, $CaCl_2$, $MnCl_2.4H_2O$, $CuSO_4.5H_2O$ and $ZnCl_2$) at 25mM concentration increases the binding in different ratios in comparison to the control as shown in figure (2-11). Cu(II) increases the binding more than other divalent cations for the three tissues homogenates the reason may be due to electrostatic interactions.

In general, the mechanism by which these salts effect protein-protein interactions is not completely clear, one hypothesis assumes that salt may alter the nature of the hydrophobic forces controlling the stabilization of protein-protein complex formed and these vary depending on the nature of the interacting groups [152]. Results illustrated in figure (2-11) suggested that these salts may provide some conformational changes in the CA125 and charge groups of the binding domain of the antibody and antigen molecules.[153]

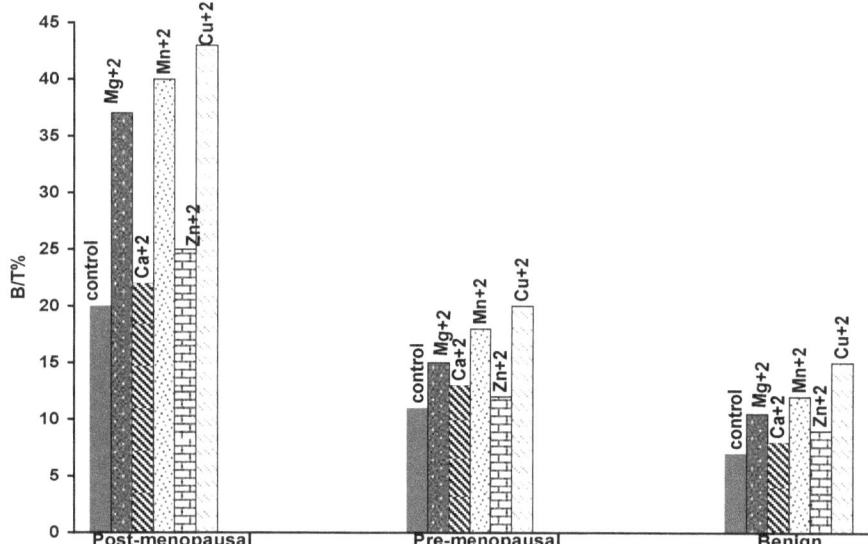

**Figure (2-11): Effect of different divalent cations on the binding of $^{125}$I-anti CA125 antibody with CA125 in ovarian tumors homogenates OI.OII and OIII (All other details are explained in the text).**

Figure (2-12) illustrated the effect of monovalent cations (KCl and NH$_4$Cl) on the binding of $^{125}$I-antiCA125 antibody to its antigen. KCl at 25 mM concentration was shown to increase the binding of the three tissues homogenates. While NH$_4$Cl at the same concentration seemed to increase the binding of post and pre-menopausal malignant ovarian tumor homogenates and slightly inhibited the binding 0f benign ovarian tumor which may be due to the presupposition that the lesser degree of hydration permits greater interaction of the salt with an ionic group located in the antibody combining site and then inhibits the complex formation. [11]

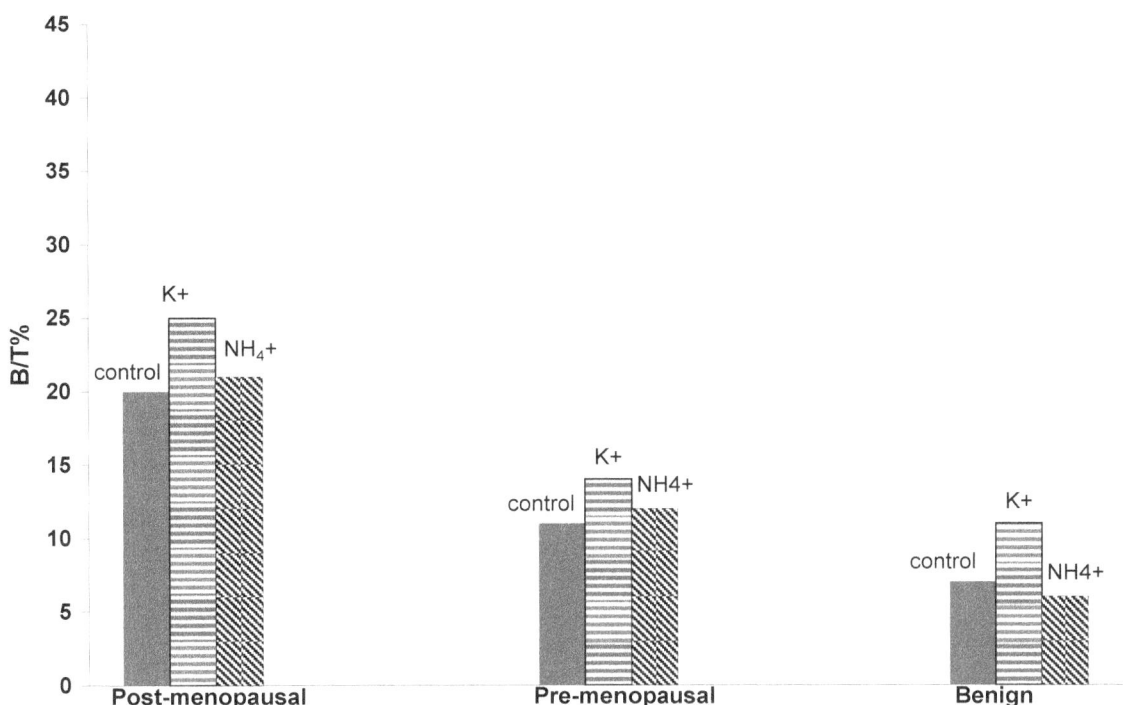

**Figure (2-12):** Effect of different monovalent cations on the binding of [125]I-antiCA125 antibody with CA125 in ovarian tumors homogenates OI, OII and OIII.

(All other details are explained in the text).

# Chapter Three

# Chromatographic purification of CA125 by Gel Filtration and Binding Characterization to Its Specific Antibody.

Abstract

Cancer antigen CA125 was partially purified from homogenate of malignant ovarian tumor by Gel filtration chromatography technique.

The results revealed   presence of two forms of CA125 antigen (BI) and (BII) with molecular weight 670 and 100 KD respectively. (BI)  form possesses a high affinity for the binding to its antibody $^{125}$I-antiCA125 in comparison to (BII) form.

The elution volume (Ve) and the $K_{av}$ values for elution of CA125 from Sepharose CL-6B column were calculated. The experiments of optimum conditions of binding between the two forms of CA125 antigen and $^{125}$I-anti CA125 antibody were determined.

## 3.1 Introduction

CA125 antigen has been characterized as a high molecular weight glycoprotein aggregate with notable size heterogeneity ranged from 200 - 1000 kDa.

Many authors tried to isolate and purify CA125 from different sources like sera of ovarian carcinoma patients [94], ovarian cancer cell line (OVCA433) [97] and human milk, using size exclusion chromatography, affinity chromatography, gel electrophoresis and buoyant density ultracentrifugation techniques [97]. Another study found that CA125 antigen isolated from amoniotic fluid using gel filtration and anion exchange chromatography was composed of two subunits of approximately 240 and 180 kDa as detected by iodine 125-lablled OC125 monoclonal antibody. [154]

In this present study, post-menopausal malignant ovarian tumor tissue was used as a source for partial purification of CA125. The factors affecting the binding of partial purified CA125 to $^{125}$I-antiCA125 antibody was also studied.

Materials and methods

## 3.2 Materials

### 3.2.1 Chemicals

All chemicals and reagents mentioned in section (2.2.1) were used in the experiments of this chapter.

### 3.2.2 Instruments

All instruments mentioned in section (2.2.2) were also used in the experiments of this chapter.

### 3.2.3 Patients

The tissues homogenates of post-menopausal patients with ovarian cancer (OI) were used in the following experiments.

## 3.3 Methods

### 3.3.1 Partial Purification of CA125 by Sepharose CL-6B Column.

### 3.3.1.1 Preparation of the Column

The dimensions of the column were chosen according to the following equation. [155]

$$\text{Diameter} = \sqrt[3]{\frac{m}{10}}$$

Where m = amount of protein in mg.

L = 30 x diameter

L = length of the column.

### 3.3.1.2 Preparation of the Buffer

Tris-HCl buffer of 0.05M, pH=7.2 containing 0.02% Sodium azide was prepared as mentioned previously in section (2.3.4).

### 3.3.1.3 Preparation of the Gel

Sepharose CL-6B gel was prepared by allowing the pre-swollen gel to swell again in Tris-buffer (0.05M pH 7.2) then left to settle and the excess of buffer was decanted. The step was repeated several times. The gel was degassed using evacuation pump and slurry was left for 24 hrs to equilibrate with buffer.[97]

The swollen gel was suspended and carefully poured into a vertical glass column (1.0x27cm) down the wall using a glass rod. After the gel has settled, the column was equilibrated with Tris-buffer for 24 hrs.

### 3.3.1.4 Void Volume Determination

The void volume of the column was determined by using blue dextran 2000 at a concentration of 2mg.ml$^{-1}$ dissolved in Tris-buffer pH 7.2, and then the elution was carried out with the same buffer at a flow rate of 12 ml /1 hr at 10°C.

Fractions of 1 ml were collected and their absorbance was measured at 600 nm. Figure (3.1) shows the elution profile of blue dextran 2000. The volume of the buffer required to elute the blue dextran, which represents the void volume, was (10ml).

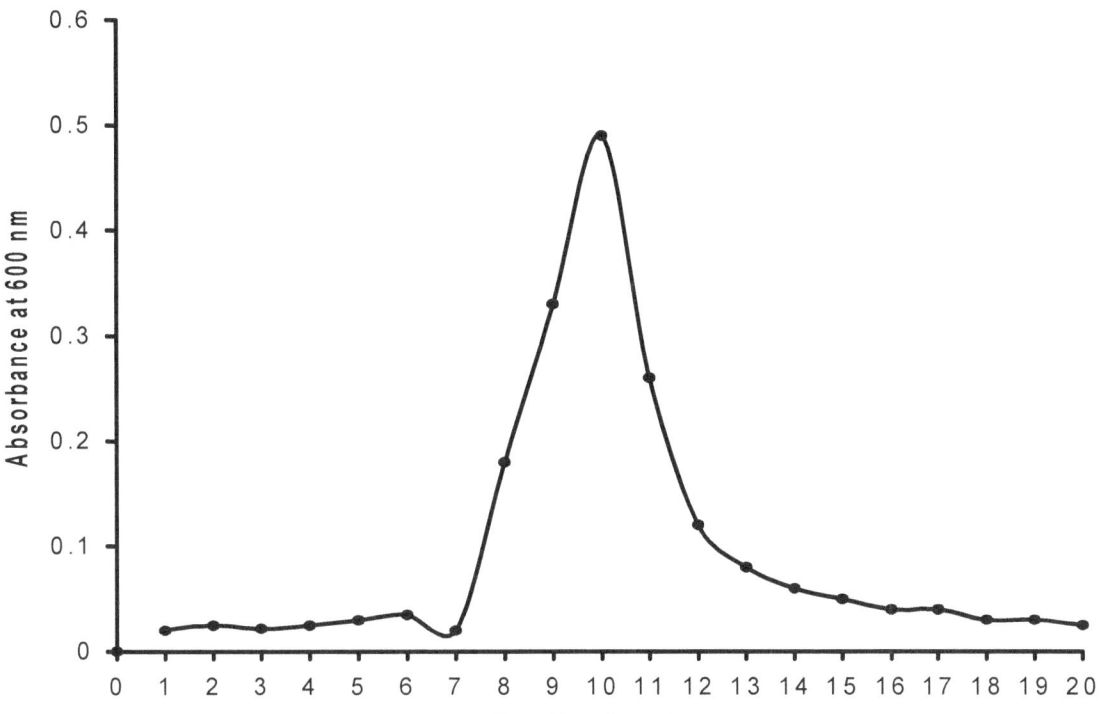

**Figure (3-1): The elution profile of blue dextran 2000 using sepharose CL-6B gel, 12 ml/hr flow rate, Tris-buffer pH 7.2 and 10°C. (All other details are explained in the text).**

### 3.3.1.5 Determination of the Molecular Weight by Gel Filtration Chromatography:

Pharmacia calibration kit for determination of M.wt by gel filtration was used. The kit comprises highly purified proteins individually packed. Each protein was reconstituted in 1.0 ml. Tris buffer pH 7.2. The standard proteins and their M.wt are detailed in table (3.1).

**Table (3-1): Standard proteins and their molecular weights (All other details are explained in the text).**

| Protein | M.wt kDa | Conc.mg.ml$^{-1}$ |
|---------|----------|-------------------|
| Thyroglobulin | 669 | 5.0 |
| Ferritin | 440 | 1.0 |
| Catalase | 232 | 5.0 |
| Aldolase | 158 | 5.0 |

## Procedure

The same sepharose CL-6B column used in   this section (3.3.1) was calibrated for molecular weight determination. Standard protein solutions were prepared according to the manufacturer instruction, then applied through two portions (0.5ml/portion), thyroglobulin and catalase in the first, Ferritin and Aldolase in the second. Elution was carried out with Tris buffer pH 7.2 at a flow rate 12 ml/hr at 10°C. The absorbencies of the fractions collected were measured at 280 nm to evaluate the elution volume (Ve) of the standard proteins.

## Calculations

The Kav of the proteins eluted were determined using the following equation [156]

$$Kav = \frac{Ve - V_O}{Vt - V_O}$$

Where:

$V_0$ = void volume

$V_e$ = Elution volume

$V_t$ = Total gel bed volume, which was calculated according o the following:

$$V_t = \left(\frac{d}{2}\right)^2 \times 3.14 \times h \qquad (d=1.0 \text{ cm}, h= 27 \text{ cm})$$

The calibration curve of Kav against log M.wt of the proteins  was plotted.

### 3.3.1.6 Separation of CA125 from Malignant postmenopausal Ovarian Tumor Homogenate

Tris-buffer (0.05M, pH7.2) containing 0.02% Sodium azide was prepared as described previously in section (2.3.4.).

## Procedure

The sample of tissue homogenate (500µl) containing approximately 7.8mg protein was applied to the surface of the gel, equilibrated with Tris-buffer 0.05M, pH 7.2 the sample was eluted using the same buffer with flow rate of 12ml/hr and fractions volume of 1ml each were collected. Gel filtration was carried out at 10°C and the absorbance of each fraction was measured at 280 nm.

The fractions that contained CA125 antigen was identified by the assay method as follows:

1- Twenty five micro liters (450 µg.ml$^{-1}$) of $^{125}$I-antiCA125 antibody was added to 100µl of each fraction number of post-menopausal malignant ovarian tumor homogenate. Then all tubes were completed to 400µl with Tris-buffer (0.05M, pH 7.2).

2- Steps 2, 3, 4, 5, and 6 mentioned in experiment (2.3.5.1) were repeated.

## Calculations

1- The absorbance of each fraction was determined at 280 nm.

2- The value of B/T ratio for the eluted fractions was calculated as mentioned in section 2.3.4.

3- The values of B/T ratio and the absorbencies at 280nm were plotted against the fraction number.

### 3.3.1.7 Dialysis for Concentration

After preparing dialysis tube, the fractions that contained high level of the binding activity were pooled and concentrated by dialyzing against sucrose at 4°C for 3hrs to get the required concentration to be used in the next experiments.

## 3.4 Determination of the optimum Reaction Conditions for the Binding of the partially purified CA125 antigen to $^{125}$I-antiCA125 antibody.

The optimum reaction conditions for the binding of partially purified CA125 to its $^{125}$I-antiCA125 antibody were studied using the same experiments mentioned for the factors affecting the binding in chapter 2.

## 3.5 Results and Discussion
### 3.5.1. Partial Purification of CA125

Partial purification of CA125 was performed by gel exclusion chromatography technique. Post-menopausal malignant ovarian tumor homogenate (OI) was applied to sepharose CL-6B (1.0x27cm) column. The void volume (Vo) of this column was (10ml) as predicted from the elution profile of the blue dextran figure (3-1).

Figure (3-2) shows the elution profile for (OI) homogenate after measuring the absorbance of collected fractions at 280 nm. It gave three main peaks separated according to their molecular weight, their fractions number were 10, 19 and 28.

The binding reaction that was carried out for the collected fractions gave two peaks (BI&BII) at fraction number 11 and the fraction number 21 for BI & BII respectively, as shown in figure (3-2). The resultant fractions containing the binding activity of CA125 were collected, pooled and concentrated then subjected to protein determination as described in section (2.3.1).

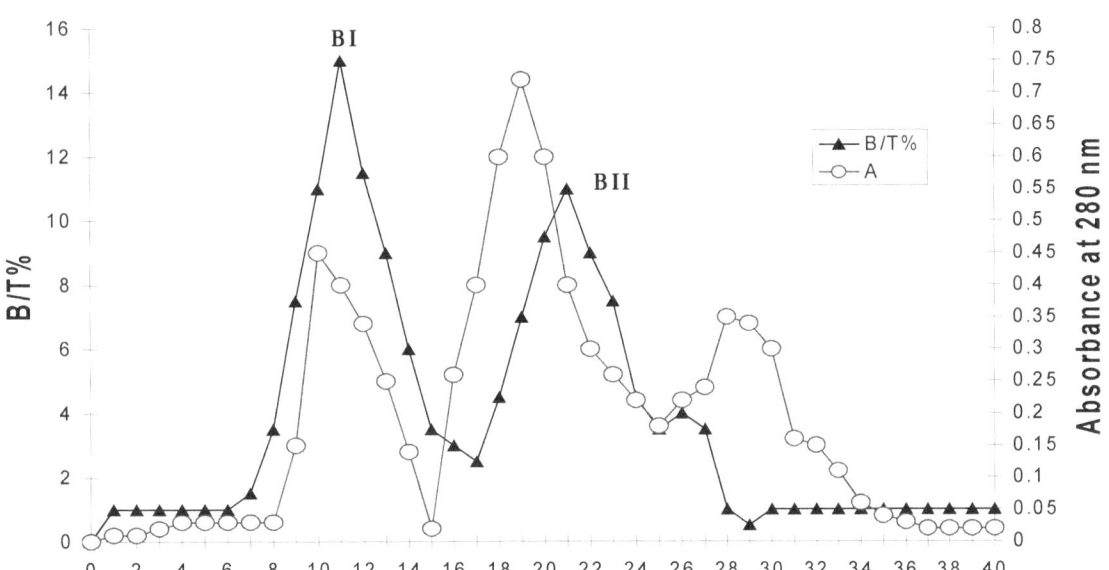

**Figure (3-2): The elution profile of human CA125 from post-menopausal malignant ovarian tumor using sepharose CL-6B gel, 12ml/hr flow rate, Tris-buffer pH 7.2, at 10°C.(All other details are explained in the text).**

89

Different standard proteins of known molecular weights were used to determine the molecular weight of the isolated antigens. The elution volumes (Ve) of standard proteins are shown in figure (3-3). The Kav values for these standard proteins were calculated by using the formula represented in section (3.3.1.4) and the calibration curve was plotted between Kav values of the standard proteins versus their logarithmic molecular weight as shown in figure (3-4).

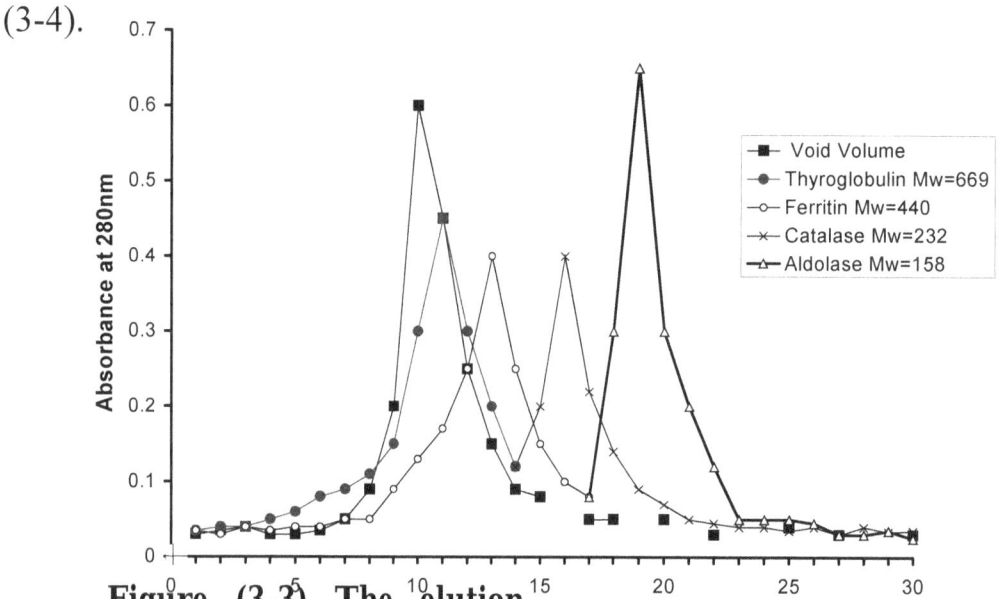

**Figure (3-3) The elution profile of standard proteins using sepharose CL-6B gel, 12 ml/hr flow rate, Tris-buffer pH 7.2, at 10°C. (All other details are explained in the text.**

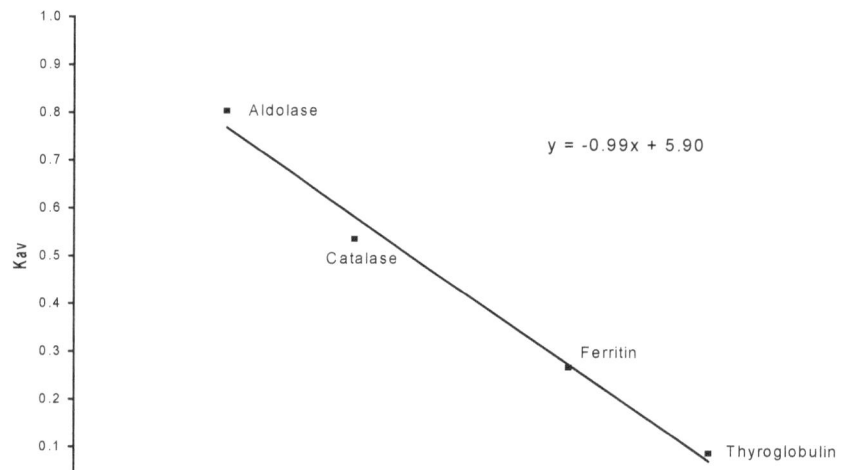

**Figure (3-4) Calibration curve for determination of M.wt by gel filtration chromatography (All other details are explained in the text).**

The straight line equation generated from this plot (figure 3.4) was used for the determination of the molecular weight of partially purified CA125. The results shows that both forms have high molecular weight where the first BI has 670 kDa and the second (BII) has 100 kDa. These results may be explained according to the idea that the lower molecular weight antigen (BII) is considered a breakdown product of a high molecular weight species and therefore contains the same antigenic determinants [154,157]. On the other hand high molecular antigen (BI) may be explained according to that CA125 has high molecular mass aggregate. The size exclusion of native CA125 from body fluids gave at least two broad peaks reactivity of 200 and > 1000kDa molecular weight. [99]

### 3.5.2 Determination of the Optimum Reaction Conditions for the Binding of Partially Purified CA125 to $^{125}$I- antiCA125 antibody

#### 3.5.2.1. Optimum protein concentration

Figure (3-5) shows the effect of increasing amount of partially purified CA125 (BI & BII) to fixed amount of $^{125}$I-anti CA125-antibody to produce ($^{125}$I-antiCA125 antibody/ CA125) complex. The shape of the curve is similar to that obtained for the crude CA125, the figure shows that the amount of partially purified CA125 needed to reach maximum binding with $^{125}$I-antiCA125-antibody as 90, 110 µg.ml$^{-1}$ which is less than the amount needed for crude extract (225 µg.ml$^{-1}$).

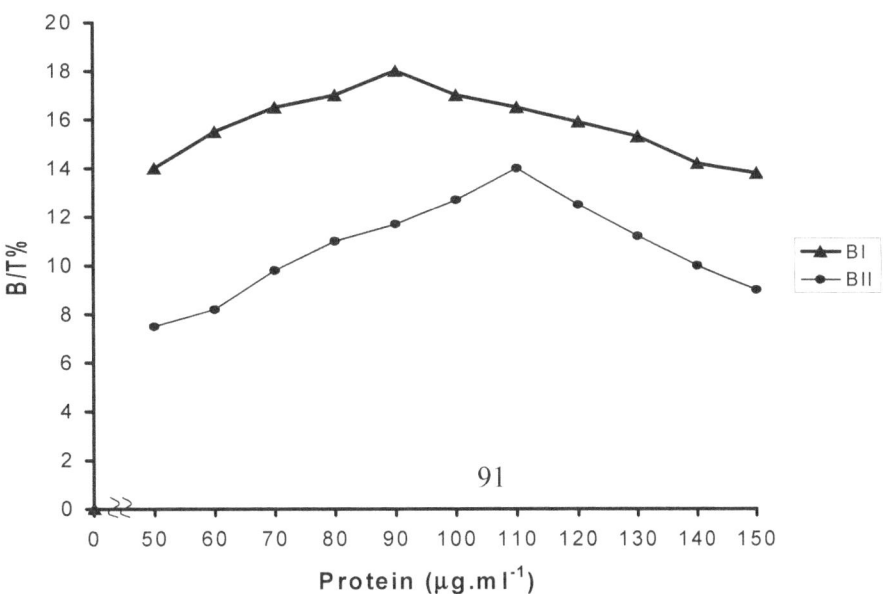

91

**Figure (3-5): The effect of protein concentration on the binding of $^{125}$I-anti CA125 antibody with partially purified CA125 (BI and BII) (All other details are explained in the text).**

### 3.5.2.2. Optimum $^{125}$I- anti CA125-antibody Concentration.

Figure (3-6) illustrates the effect of increasing $^{125}$I-antiCA125 antibody concentration on the binding with partially purified forms of CA125 (BI & BII). The maximum binding obtained when 360 µg.ml$^{-1}$ for (BI) and 450 µg.ml$^{-1}$ for (BII) were used. From these result it was found that partially purified CA125 (BI) form was saturated with small concentration of $^{125}$I-antiCA125 antibody than those required for BII. Thus it was concluded that BI has higher affinities at low concentrations toward $^{125}$I-antiCA125 antibody than BII.

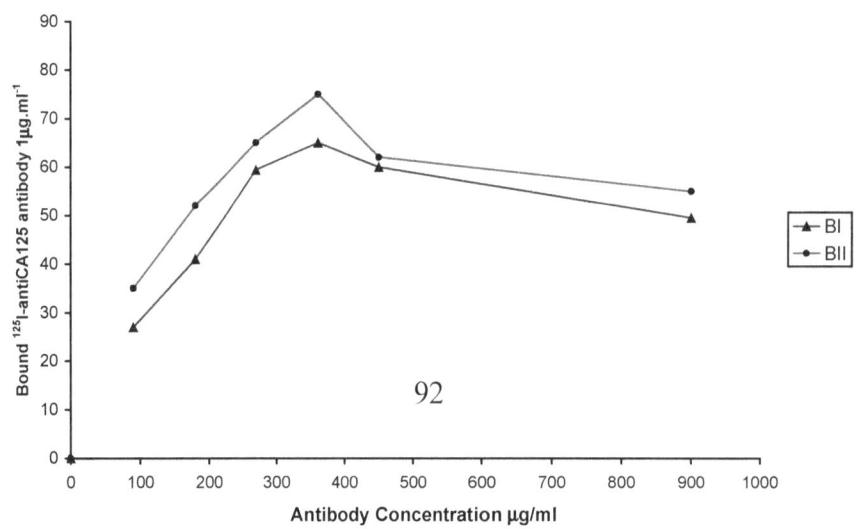

**Figure (3-6): The effect of $^{125}$I-antiCA125 antibody concentration on the binding with partially purified CA125 (BI and BII)(All other details are explained in the text).**

### 3.5.2.3. Optimum pH

Figure (3-7) shows the effect of pH on binding of $^{125}$I-antiCA125 antibody to partially purified CA125 (BI&BII) of post-menopausal malignant ovarian tumor homogenate. The results revealed that the optimum pH for (BI) and (BII) was 6.8 and 7.2 respectively. These results indicate that the binding was pH dependent and the differences in the optimum pHs may suggest the differences in the binding sites of these partially purified antigens [158].

**Figure (3-7): The effect of pH on the binding of $^{125}$I-antiCA125 antibody with partially purified CA125 (BI and BII)**
**(All other details are explained in the text).**
*3.5.2.4 The time course of the binding of partially purified CA125 (BI &BII)*

*to $^{125}$I- antiCA125 antibody.*

The optimum time and temperature for partially purified CA125 (BI and BII) was studied. Figures (3-8) and (3-9) show that the (BI) form antigen binds to its specific antibody in highest state after 180 min at 5°C, while BII form reach the maximum binding after 210 min at 5°C .

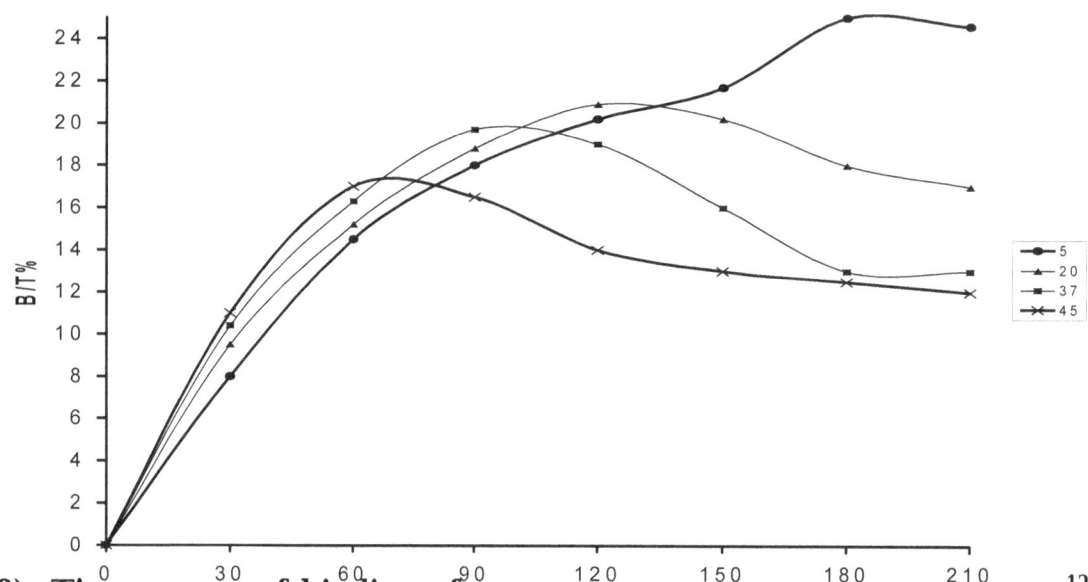

**Figure (3-8): Time course of binding of** $^{125}$I-antiCA125 **antibody with partially purified CA125 (BI) (All details are explained in the text).**

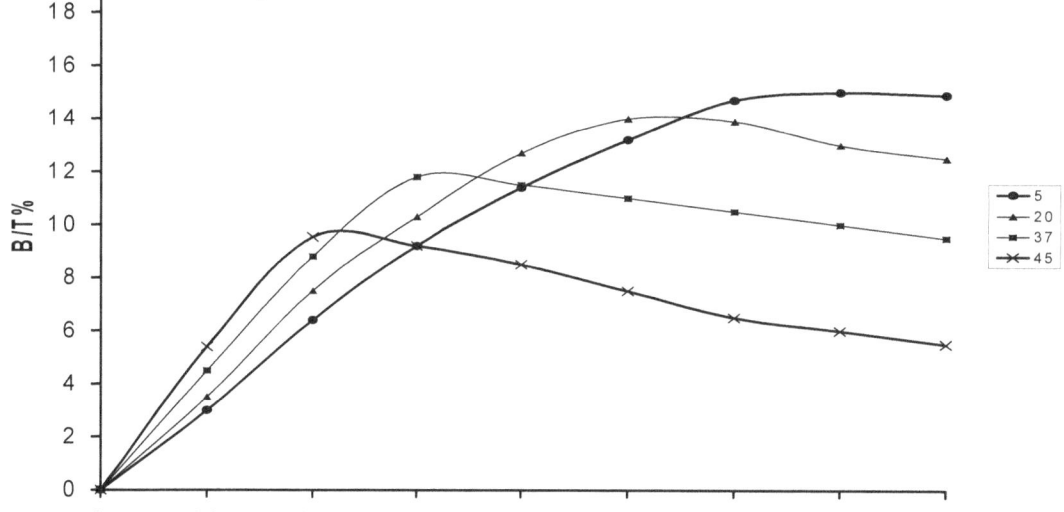

**Figure(3-9): Time course of binding of** $^{125}$I-antiCA125 **antibody with partially purified CA125 (BII)(All details are explained in the text).**

In comparison with crude homogenate, the time was shortened for both BI & BII from 240 minutes for the crude homogenate to 180 and 210 minutes for BI & BII respectively. A possible explanation for the fast kinetics of complex formation with partial purified antigen may be due to losing of inhibitory factors, by exclusive gel filtration process. This in turn affects the rate of complex formation between CA125 and its specific antibody

# Chapter Four

# Kinetic & Thermodynamic Studies of the Binding of CA125 with $^{125}$I-antiCA125 antibody in Ovarian Tumor Homogenates

# Abstract

Kinetic and thermodynamic parameters associated with the binding of [125]I-anti125 antibody to crude CA125 of benign, post and pre-menopausal malignant ovarian tumors and with partially purified CA125 (BI) and (BII) forms of post-menopausal malignant ovarian tumor were carried out using different rate equations. Then the order of the reactions were tested. It was showed that the reaction in all groups follow second rate law. In addition, the kinetic parameters $K_a$, $K_d$, $k_{+1}$, $k_{-1}$, $t_{1/2}$ of association and dissociation were determined at 5, 20, 37 and 45°C. In all studied groups the affinity constant ($K_a$) and maximum binding capacity ($B_{max}$) decrease and their rate constant ($k_{+1}$) values increase with increasing temperature. Scatchared plots of all groups showed no curvature in the plotted lines, where the data obeyed the straight line equation suggested that the CA125 has a single binding site or more than one site with identical affinities. The thermodynamics of the binding of [125]I-antiCA125 antibody with CA125 for all groups were studied using Van't Hoff and Arrhenius equations, the thermodynamic parameters of standard state ($\Delta H^o$, $\Delta G^o$ and $\Delta S^o$) were determined. The data showed that the binding reactions were exothermic ($\Delta H^o < 0$) and spontaneous ($\Delta G^o < 0$) and the binding reactions were entropically and enthalpically driven ($\Delta S^o > 0$ and $\Delta H^o < 0$). Arrhenius plot indicated that there was a linear- relationship between log $k_{+1}$ and 1/T. The transition state thermodynamic parameters (Ea, $\Delta H^*$, $\Delta G^*$ and $\Delta S^*$) for the formation of ([125]I-anti CA125 antibody/CA125) were determined.

## 4.1 Introduction

Molecular interactions involve cooperative, independent and contiguous binding regions. Molecular interaction generates dynamic structural changes, which increase the complexity of these interactions. The rate of the reaction and equilibrium conditions is the algebraic sum of the energies involved in reversible macromolecular interactions [160]. The specific reaction between antibody and antigen, as a type of these interactions, is usually driven by electrostatic forces between oppositely charged amino acids, hydrogen bonding, and hydrophobic interactions. The equilibrium reaction termed "biospecific interaction" is characterized by the affinity of the reaction to form Antibody-Antigen complex. [161]

The analysis of temperature dependence of kinetic and equilibrium constants allows determination of the energetic of binding. The changes in Gibes free energy, enthalpy and entropy that are associated with the binding of antibody to its antigen can be calculated mathematically using such constant [162].

In other words, thermodynamic measurements of reactions interactions under equilibrium conditions provide information about differences between the initial and final states of each reactant, while kinetics studies supplement the information on the differences between these states and an intermediate activated complex state, (i.e. the pathway taken by the reactants to reach the final product). [163,164]

The elucidation of biomolecular interactions is of steadily increasing importance. Exact knowledge of the principles governing the strengths and formation of molecular interaction is of highest importance for a huge number of applications in widely different areas, such as : The design of new drugs, the understanding of cross-reactions of antibodies used in medical diagnosis or medical treatment, the improvement of our understanding in diseases, the understanding and regulation of biocatalytic activity, the understanding cell-cell communication and cell differentiation, ...etc.

In this chapter, the basic mathematical analysis was described and used to explain the mechanism, through kinetics and thermodynamic, of binding of CA125 antibody to its CA125 antigen in benign, post-and pre-menopausal malignant ovarian tumor homogenates and to its partially purified forms (BI) and (BII) to form ($^{125}$I-antiCA125 antibody/CA125) complex.

Materials and Methods

## 4.2 Materials

### *4.2.1 Chemicals*

All chemicals and reagents mentioned in section (2.2.1) were used in the experiments of this chapter.

### *4.2.2 Instruments*

All instruments mentioned in section (2.2.2) were also used in the experiments of this chapter.

## 4.3 Methods

### *4.3.1. Kinetic Studies*

### *4.3.1.1 Time course of Binding*

### *A. The time course of the binding of $^{125}$I-antiCA125 antibody with CA125 in ovarian tumor homogenate*

### Reagents

All reagents were prepared according to section (2.3.4) except 20% PEG6000 which was prepared according to the optimum pH of each group.

### Procedure

1.  Tracer (450, 360and 450) µg.ml$^{-1}$ were added to (225,150 and 175 µg.ml$^{-1}$ protein) of OI, OII and OIII respectively.

2.  Each mixture was completed to 400µl with Tris-buffer at the optimum pH of each group.

3.  All tubes were incubated at 5°C at different time intervals (30, 60, 90, 120, 150, 210, 240, 270 and 300) minutes.

4.  Steps 3, 4, 5, and 6 in the section (2.3.5.1.) were repeated.

5.  To determine the time course of CA125 binding to $^{125}$I-antiCA125 antibody at different temperatures. Steps 1, 2, 3 and, 4 in this experiment were repeated at (20, 37 and 45°C).

## Calculations

The B/T% values were calculated as described in section (2.3.4.) and plotted against incubation time at different temperatures.

***B. The time course of binding of $^{125}$I-antiCA125 antibody with partially purified forms (BI) and BII of malignant ovarian tumor.***

## Reagent

All reagents were prepared according to the section (2.3.3.) except for 20% PEG 6000 which was prepared according to optimum pH of each group.

## Procedure

1. Tracer (360 and 450) $\mu g.ml^{-1}$ were added to 90 and 110 $\mu g.ml^{-1}$ protein) of BI and BII respectively.

2. Each mixture was completed to 400$\mu$l with Tris-buffer at optimum pH of each group.

3. The ($^{125}$I-antiCA125 antibody/CA125) complex for each group were estimated by following steps 3 and 4 in section 4.3.1.1.A

4. The experiment was repeated at different temperatures (20, 37 and 45°C).

## Calculations

The B/T% values were calculated as described in section (2.3.4) and plotted against incubation time at different temperatures.

***4.3.1.2. Determination of Kinetic parameters***

***A. Determination of the affinity constant ($K_a$) and maximal binding capacity ($B_{max}$) of $^{125}$I-antiCA125 antibody associated with CA125 in ovarian tumor homogenates.***

1.	Increasing volumes (4, 8, 12, 16 and 20 µl) of $^{125}$I-antiCA125 antibody containing (72, 144, 216, 288, and 360 µg.ml$^{-1}$ protein) respectively was added to each (150 µg.ml$^{-1}$) of pre-menopausal malignant ovarian tumor homogenate. Then the final reaction volume was completed to 400µl with the same buffer.

2.	All tubes were incubated at 5°C for 240 minutes.

3.	The ($^{125}$I-antiCA125 antibody/CA125) complex was estimated by following the steps 3, 4, 5, and 6 in the section (2.3.5.1.)

4.	The previous steps were performed at different temperature (20, 37 and 45°C).

5.	The experiment was repeated using increasing volumes (5, 10, 15, 20 and 25µl) of $^{125}$I-antiCA125 antibody containing (90, 180, 270, 360 and 450) µg.ml$^{-1}$ protein respectively, were added to (225 and 175) µg.ml$^{-1}$ of each of the post-menopausal malignant ovarian tumor homogenate and benign ovarian tumor homogenate respectively, instead of pre-menopausal ovarian tumor homogenate (in step 1 above). Tris-buffer at optimum pH of each group was used to complete the volume of the reaction and to prepare 20% PEG 6000.

6.	The times of incubation needed to get the equilibrium state for all cases are reported in table (4-1) .

**Table (4-1):	The time of incubation for Benign and malignant post and pre-menopausal ovarian tumor homogenates at different temperature.**

| Temp. °C | Time | | |
|---|---|---|---|
| | Post. Menopausal malignant ovarian tumor homogenate | Pre-menopausal malignant ovarian tumor homogenate | Benign ovarian tumor homogenate |
| 5 | 240 | 240 | 270 |
| 20 | 180 | 150 | 180 |
| 37 | 120 | 90 | 120 |
| 45 | 90 | 60 | 60 |

## Calculations

1.  The B/F ratio was computed for each tube, where:

    B: is the bound radioactivity (mean counts in c.p.m.) which represent the formation of ($^{125}$I-antiCA125 antibody/CA125) complex.

    F: is the free radioactivity (mean counts in c.p.m.), which represents the (unbound or unreacted), $^{125}$I-antiCA125 antibody.

    T : is the total activity (mean counts in c.p.m.)

    F : T (total counts)-B (bound radioactivity)

2.  The concentration of ($^{125}$I-antiCA125 antibody/CA125) complex in mg.ml$^{-1}$ which found after time (t) was calculated from the following equation:

    $$B \text{ (mg.ml}^{-1}) = \frac{B(c.p.m.)}{T(c.p.m.)} \times \text{concentration of } ^{125}I \text{ - antiCA125 antibody}$$ in incubation medium in mg.ml$^{-1}$

3.  The affinity constant and maximal binding capacity was determined according to scatchared equation [165].

    $$\frac{B}{F} = \frac{1}{K_d} \text{ X } (B_{max}-B)$$

    $$Ka = \frac{1}{K_d}$$

    Where $K_a$ = affinity constant

    $K_d$ = dissociation constant

    $B_{max}$ = maximal binding capacity

4.  The plot of B/F ratio Vs. the B values in mg.ml$^{-1}$ gives a linear relationship. The value of affinity constant of the binding ($K_a$) at each temperature can be calculated from the slope of the straight line, while the value of the total concentration of CA125 ($B_{max}$) in ovarian tumor homogenate of each group was calculated from the intercept of the x-axis.

### B. Determination of the affinity constant ($K_a$) and maximal binding capacity ($B_{max}$) of $^{125}$I-antiCA125 antibody associated with partially purified CA125 (BI) and (BII) of post-menopausal ovarian tumor homogenate.

1.  Increasing volume (4, 8, 12, 16 and 20 µl) of $^{125}$I-antiCA125 antibody containing (72,144,216,288 and 360) µg.ml$^{-1}$ protein respectively were each added to (90 µg.ml$^{-1}$) of partially purified CA125 (BI) form. Then final reaction volume was completed to 400 µl with the same buffer.

2.  All tubes were incubated at 5°C for 180 minute

3.  The ($^{125}$I-antiCA125 antibody/CA125) complex was estimated by following the steps 3, 4, 5, and 6 in the section (2.3.5.1).

4.  The previous steps were performed at different temperature (20, 37 and 45°C).

5.  The experiment was repeated using increasing volumes (5, 10, 15, 20 and 25µl) of $^{125}$I-antiCA125 antibody containing (90,180,270,360 and 450µg protein) respectively were each added to (110 µg) of partially purified CA125 of post-menopausal ovarian tumor homogenate (BII) form instead of (BI) (in step 1 above). Tris-buffer at pH = 6.8 was used to complete the volume of the reaction and to prepare 20% PEG6000.

6.  The times of incubation needed to get the equilibrium state for both cases are reported in table (4-2) .

**Table (4-2): The time of incubation for partially purified CA125 of postmenopausal ovarian tumor homogenate at different temperatures.**

| Temp.°C | Time min. | |
| | Partially purified CA125 (BI) form | Partially purified CA125 (BII) form |
| --- | --- | --- |
| 5 | 180 | 210 |
| 20 | 120 | 150 |
| 37 | 90 | 90 |
| 45 | 60 | 60 |

### Calculations

The method outlined in experiment (4.3.1.2A) was followed exactly to out line the values of $K_a$ and $B_{max}$ at each temperature.


## 4.3.2 The thermodynamic Studies

### 4.3.2.1 The thermodynamic studies of the interaction of [125]I-antiCA125 antibody with CA125 in ovarian tumor homogenates.

The same steps mentioned in section (4.3.1.1. A) and (4.3.1.2. A) were performed using protein fraction of Benign, malignant post-menopausal and pre-menopausal ovarian tumor homogenates.


## Calculations

1. The thermodynamic parameters of standard state were obtained from Van't Hoff plot, the values of the natural logarithm of equilibrium constant (affinity constant $K_a$) obtained at different temperatures were plotted against the reciprocal values of the absolute temperature in kelvin (1/T), according to the following equation:

$$\ln K_a = \frac{\Delta S^\circ}{R} - \frac{\Delta H^\circ}{RT}$$

**Where:**

$\Delta H^\circ$ = The enthalpy change of the standard state.

$\Delta S^\circ$ = The entropy change of the standard state.

R    = The gas constant ($8.314$ J.K$^{-1}$.mol$^{-1}$)

$\Delta H$ value was obtained from the slope of the linear relationship of the plot.

The change in Gibbs free energy of the standard state ($\Delta G^\circ$) was obtained from the following equation:

$\Delta G^\circ = -RT \ln K_a$

where $K_a$ is the affinity constant, while the standard state entropy ($\Delta S^\circ$) change was obtained from.

$$\Delta S^\circ = \frac{\Delta H^\circ - \Delta G^\circ}{T}$$

2.     The thermodynamic parameters of the transition state were obtained from Arrhenius plot of $\ln k_{+1}$ values against $(1/T)$ values, that given a linear relationship according to the following equation

$$\ln k_{+1} = \ln A - \left[\frac{Ea}{RT}\right]$$

Where:

A: Arrhenius constant.

The value of the activation energy (Ea) of the binding reaction can be determined from the slope of the straight line.

The enthalpy of transition state $\Delta H^*$ was obtained from

$$\Delta H^* = Ea - RT$$

Transition state of free energy change $\Delta G^*$ is calculated from the following equation

$$\Delta G^* = -RT \ln k_{+1} + RT \ln \frac{KT}{h}$$

where K and h were boltzman and Blank's constants which are equal to $(1.38 \times 10^{-23}\ J.K^{-1})$, $(6.62 \times 10^{-34}\ J.sec^{-1})$ respectively.

The change in entropy of the transition state $\Delta S^*$ was calculated from the following equation:

$$\Delta S^* = \frac{\Delta H^* - \Delta G^*}{T}$$

### 4.3.2.2 The thermodynamic studies of the interaction of [125]I-antiCA125 antibody with partially purified CA125 (BI) and (BII) of post-menopausal ovarian tumor homogenate.

The experiment was performed as described in section (4.3.1.1.B) and (4.3.1.2.B) using partially purified CA125 (BI) and (BII) forms of malignant post – menopausal ovarian tumor homogenate.

### Calculations

The method outlined in the experiment (4.3.2.1.) was followed exactly for estimating the thermodynamic parameters of the standard and transition state.

## 4.4. Results and Discussion

### 4.4.1 Determination of kinetic parameters of CA125 Associated with $^{125}I$-antiCA125 antibody.

The time course of $(^{125}I$-antiCA125 antibody/CA125) complex formation was carried out to describe the kinetic parameters of the binding. The simplest proposed model representing this interaction is:

$$^{125}\text{I-antiCA125 antibody} + \text{CA125} \underset{k_{-1}}{\overset{k_{+1}}{\rightleftharpoons}} [^{125}\text{I-antiCA125 antibody/CA125}]$$

Where

$k_{+1}$: is the association rate of $^{125}I$-antiCA125 antibody to CA125.

$k_{-1}$: is the dissociation rate of $^{125}I$-antiCA125 antibody/CA125) complex formed.

At equilibrium

$$K_a = \frac{\left[^{125}\text{I - antiCA125 antibody / CA125}\right]}{\left[^{125}\text{I - antiCA125 antibody}\right]\left[\text{CA 125}\right]} \quad \cdots\cdots (1)$$

$$K_d = \frac{\left[^{125}\text{I - antiCA125 antibody}\right]\left[\text{CA 125}\right]}{\left[^{125}\text{I - antiCA125 antibody / CA125}\right]} \quad \cdots\cdots (2)$$

Thus:

$$K_a = \frac{1}{K_d} = \frac{k_{+1}}{k_{-1}} \quad \cdots\cdots (3)$$

The value of $K_a$ and maximal binding capacity ($B_{max}$) were calculated from scatchared plot at four different temperatures for all studied groups [OI, OII, OIII and partially purified CA125 (BI & BII) of malignant post-menopausal ovarian tumor homogenate. The experiment was carried out at the optimum conditions that were obtained in previous experiments and was repeated at different temperatures (20, 37 and 45°C).

Scatchared plots were analyzed according to their linearity as shown in figures (4.1 A, B and C) and (4.2.A & B), all groups showed no curvature in the plotted lines where the data obeyed the straight line equation, indicating that CA125 has a single binding site or more than one site with identical affinities.

Table (4.3) shows that the affinity constant ($K_a$) and ($B_{max}$) depended on the type of tumor (benign or malignant) and on the temperature. $K_a$ decreases with the increased temperature in all studied groups. The highest value of $K^a$ occurred in OI group at 5°C, it is about (5.432 mg.ml$^{-1}$) which suggest the highest affinity for binding among the two rest groups. The increase in temperature may affect the protein conformation which leads to decrease the affinity of binding. On the other hand, determination of ($B_{max}$) of CA125 to each type of tissue homogenate shows similar result for $K_a$ value, it is temperature depended, $B_{max}$ decreased with increasing temperature.

The results in table (4-4) also reveal that there is a decrease in $K_a$ and $B_{max}$ values for partially purified CA125 (BI) and (BII) forms with increasing temperature. $K_a$ for BI and BII was (7.092 mg$^{-1}$.ml and 3.333mg$^{-1}$.ml at 5 °C respectively, while it was 3.20 mg$^{-1}$.ml and 2.080 mg.ml$^{-1}$ at 45°C for the two groups respectively. In comparison of $K_a$ and $B_{max}$ values for BI and BII, BI show higher affinity and lower binding capacity than BII.

In general, it can be concluded that partially purified CA125 (BI) form interacts with its specific antibody with higher affinity than the interaction of crude CA125 antigen.

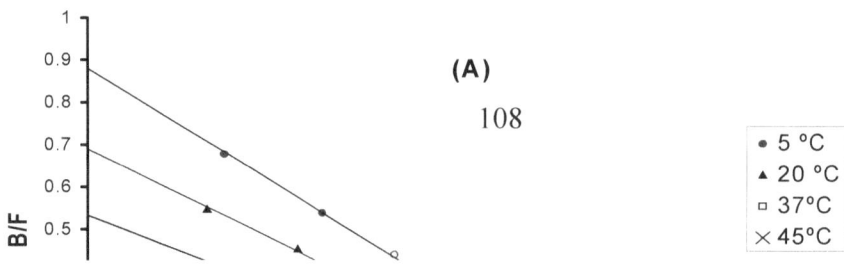

**(A)**

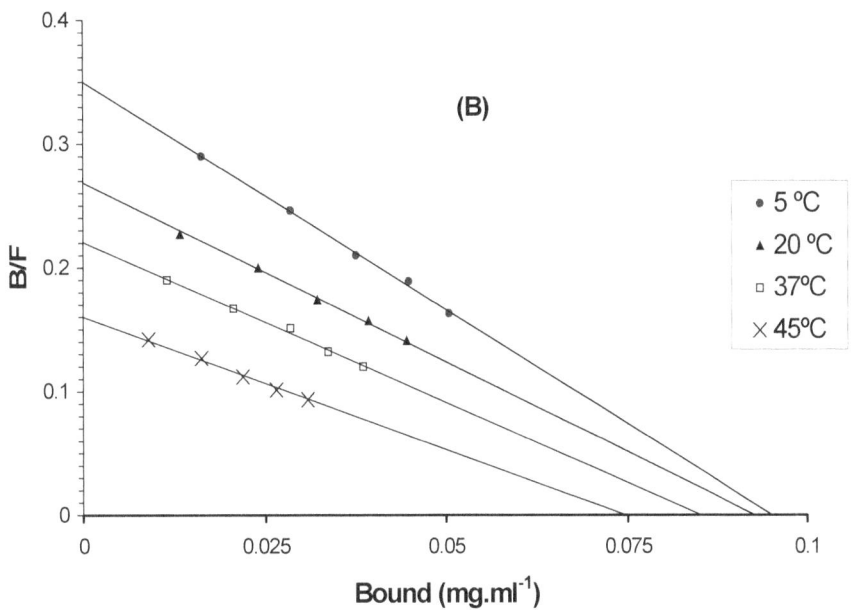

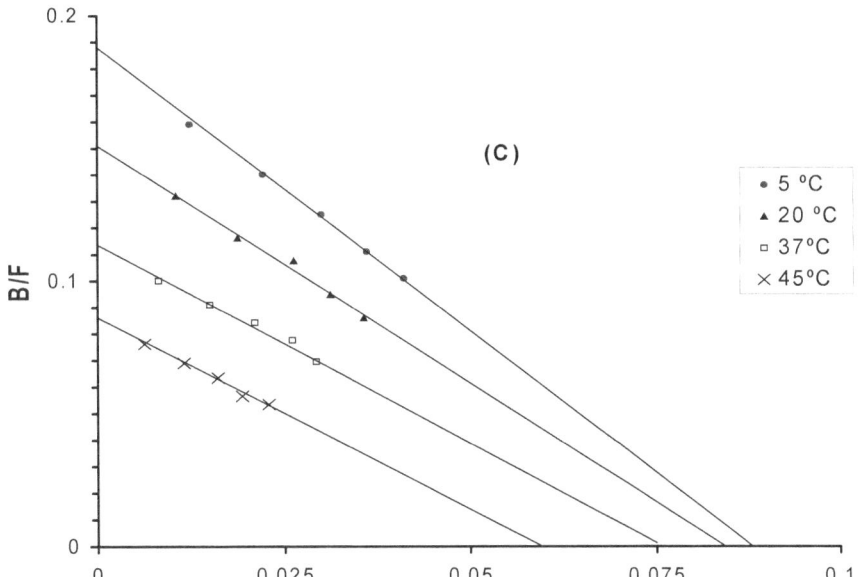

**Figure(4-1): Scatchard Plot for A)OI, B)OII, C) OIII at different temperatures. (All other details are explained in the text).**

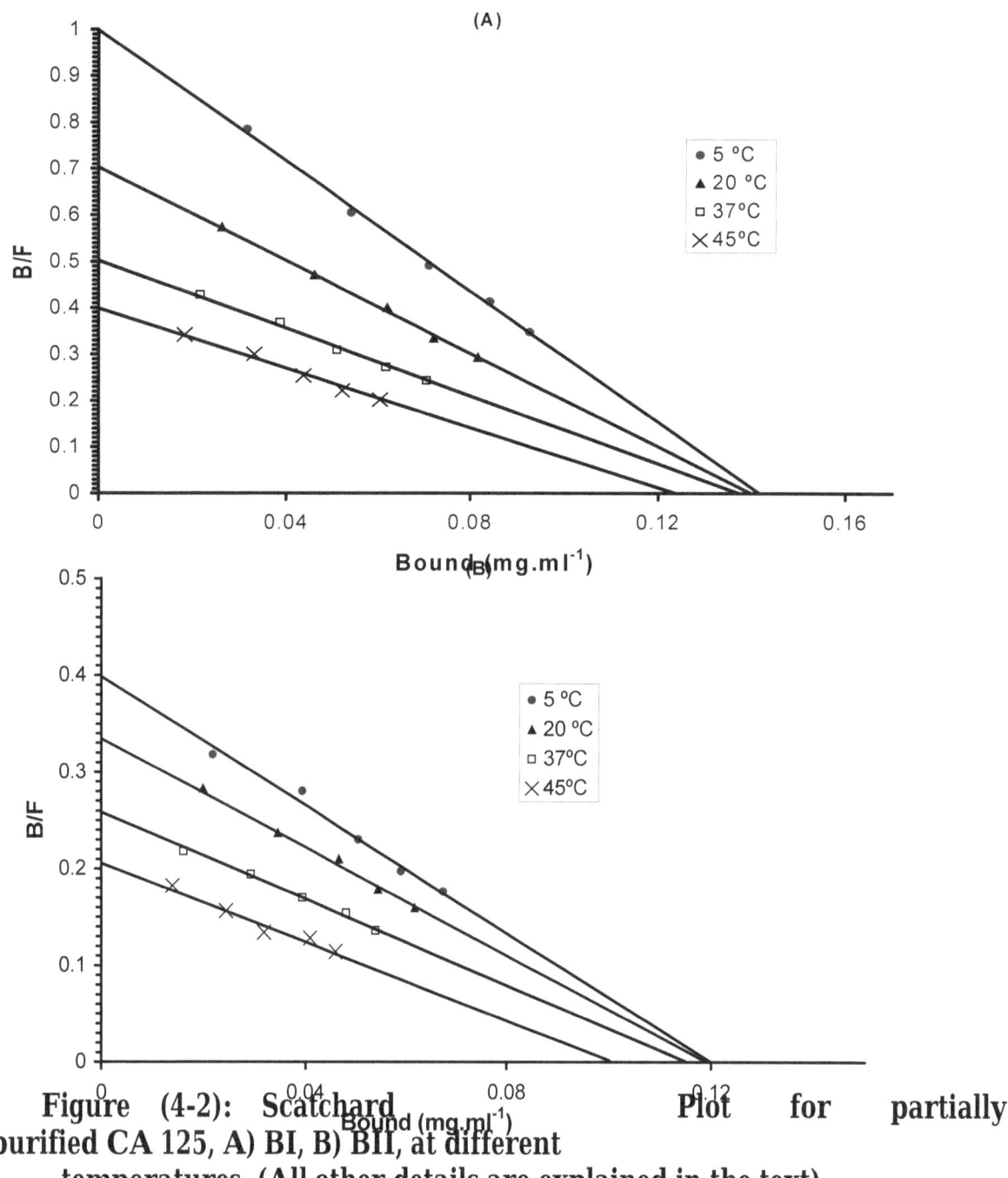

Figure (4-2): Scatchard Plot for partially purified CA 125, A) BI, B) BII, at different temperatures. (All other details are explained in the text).

**Table 4-3:** Association, dissociation constants and maximal binding capacity of the binding of $^{125}$I-antibody CA125 antibody to CA125 antigen in ovarian tumor homogenates at different temperatures

| Temperature °C | $K_a$ (mg$^{-1}$.ml) | $K_d$ (mg.ml$^{-1}$) | $B_{max}$ (mg.ml$^{-1}$) |
|---|---|---|---|
| Malignant post-menopausal ovarian tumor | | | |
| 5 | 5.432 | 0.184 | 0.162 |
| 20 | 4.367 | 0.228 | 0.158 |
| 37 | 3.419 | 0.292 | 0.155 |
| 45 | 3.034 | 0.329 | 0.145 |
| Malignant pre-menopausal ovarian tumor | | | |
| 2 | 3.684 | 0.271 | 0.095 |
| 20 | 3.027 | 0.330 | 0.092 |
| 37 | 2.588 | 0.386 | 0.085 |
| 45 | 2.133 | 0.468 | 0.075 |
| Benign ovarian tumor | | | |
| 5 | 2.148 | 0.465 | 0.087 |
| 20 | 1.800 | 0.555 | 0.083 |
| 37 | 1.554 | 0.643 | 0.074 |
| 45 | 1.433 | 0.697 | 0.060 |

**Table (4-4):** Association, dissociation constants and maximal binding capacity of the binding of $^{125}$I-antiCA125 antibody to partially purified CA125 BI and BII of malignant post-menopausal ovarian tumor homogenates at different temperatures.

| Temperature °C | $K_a$ mg$^{-1}$.ml | $K_d$ (mg.ml$^{-1}$) | $B_{max}$ (mg.ml$^{-1}$) |
|---|---|---|---|
| | | | Partially purified (BI) |
| 5 | 7.092 | 0.141 | 0.141 |
| 20 | 5.000 | 0.200 | 0.140 |
| 37 | 3.676 | 0.272 | 0.137 |
| 45 | 3.2 | 0.312 | 0.125 |
| Partially purified (BII) | | | |
| 2 | 3.333 | 0.300 | 0.120 |
| 20 | 2.750 | 0.363 | 0.120 |
| 37 | 2.263 | 0.441 | 0.110 |
| 45 | 2.080 | 0.480 | 0.100 |

However the time course data could be used to determine the reaction order of CA125 binding to its specific antibody using the graphical method. Attempts were carried out using pseudo first order and second order graphs.

For second order graph the following equations [163] were used

$$\ln[AbAg]_e \left[ \frac{[Ab]_T - [AbAg]_t [AbAg]_e / [Ag]_T}{[Ab]_T ([AbAg]_e - [AbAg]_t)} \right] = k_{+1}t \left[ \frac{[Ab]_T [Ag]_T - [AbAg]_e}{[AbAg]_e} \right] \ldots 4$$

Where:

$K_{+1}$: is the association rate constant.

$[AbAg]_e$: is the concentration of ($^{125}$I-antibodyCA125 antibody/CA125)complex formed at equilibrium.

$[AbAg]_t$: is the concentration of ($^{125}$I-antibodyCA125 antibody/CA125)complex after time t.

$[Ab]_T$: is the initial antibody concentration at time 0.

$[Ag]_T$: is the initial antigen concentration at time 0.

Or by using another second order kinetic equation from:

$$\frac{1}{[Ab]_T - [Ag]_T} . \ln\left( \frac{[Ab]_T - [AbAg]_t}{[Ag]_T - [AbAg]_t} \right) = K_{+1}t + \frac{1}{[Ab]_T - [Ag]_T} \ln \frac{[Ab]_T}{[Ag]_T} \ldots 5$$


For first order graph as the percent of binding was in some cases small [166] and must be labeled antibody remains free and only small fractions binds even at equilibrium ,ie. $[Ab]_T >> [AbAg]_e$

Thus

$$[Ab]_T >> \frac{[AbAg]_t [AbAg]_e}{[Ag]_T}$$

So the following equation could be used in order to fit the pseudo-first order kinetics:

$$\ln \frac{[AbAg]_e}{[AbAg]_e - [AbAg]_t} = K_{+1}.t \frac{[Ab]_T [Ag]_T}{[AbAg]_e} \ldots\ldots 6$$

On the other hand figure (4-3 A, B &C) and (4-4 A&B) show the plot of

$$\frac{1}{[Ab]_T - [Ag]_T} \ln\left( \frac{[Ab]_T - [AbAg]_t}{[Ag]_T - [AbAg]_t} \right)$$

Against time give a straight line for all studies groups OI, OII, OIII and partially purified CA125 (BI and BII form).The association rate constant $k_{+1}$ was determined at each temperature from the slope of the plot.

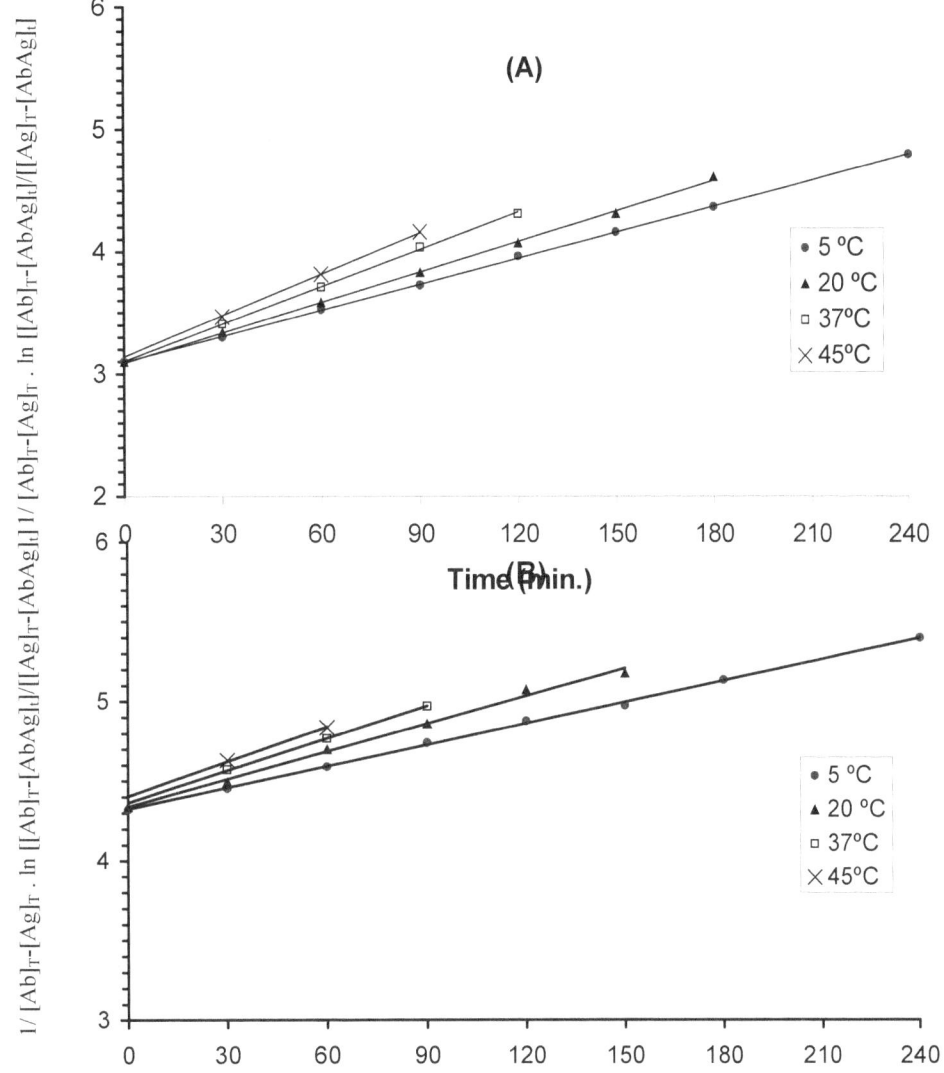

**Figure (4-3): Kinetics of the binding of $^{125}$I-anti CA125 antibody to CA125 of**

**A: Malignant post-menopausal ovarian tumor homogenate.**

**B: Malignant pre-menopausal ovarian tumor homogenate.**

**C: Benign ovarian tumor homogenate. At different temperature using second order rate law (All other details are explained in the text).**

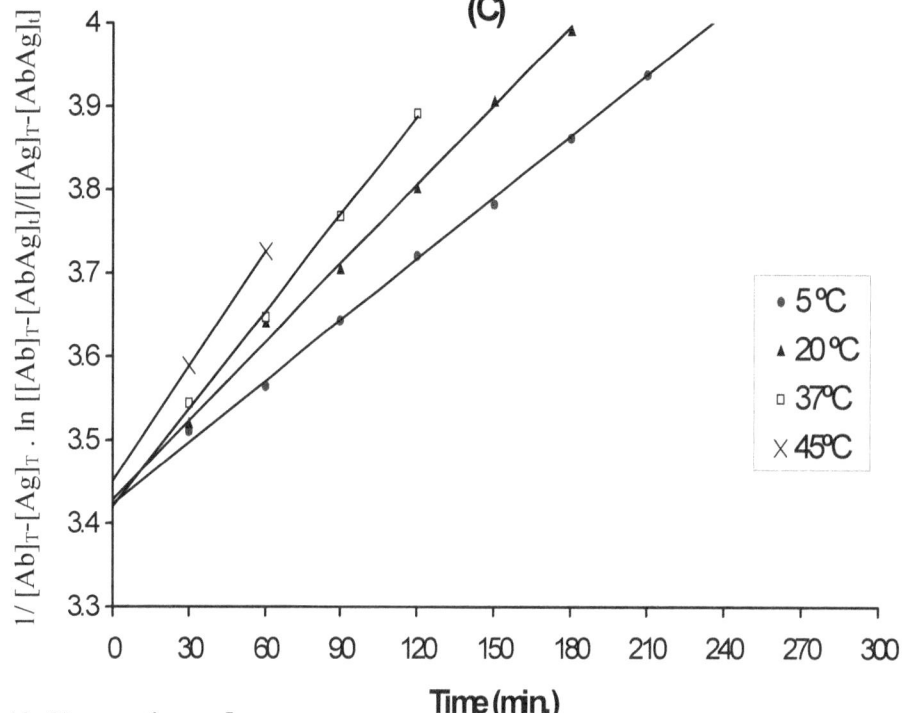

**Figure (4-3) continued**

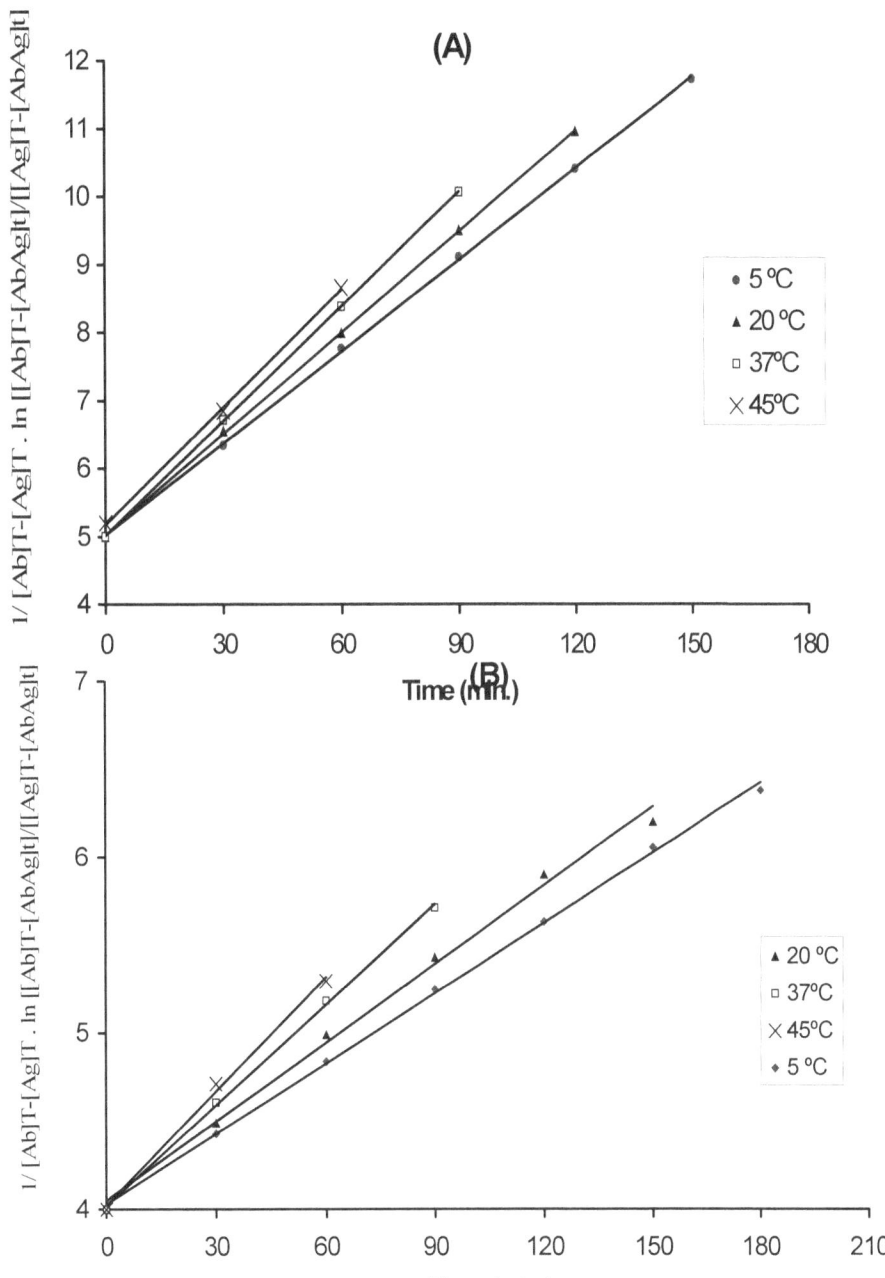

**Figure (4-4): Kinetics of the binding of $^{125}$I-anti CA125 antibody with partially purified CA125 BI and BII from malignant post- menopausal ovarian tumor homogenate. At different temperature using second order rate law (All other details are explained in the text).**

Kinetic parameters for all studied groups (OI, OII, OIII, BI and BII) were illustrated in tables (4-5) and (4-6). The value of $k_{+1}$ indicates that highest rate of association of CA125 with $^{125}$I-antiCA125 antibody occurs at 45°C, whereas the lowest rate occurs at 5°C. Thus when the reaction temperature was increased

115

from 5 °C to 45°C, the values of the association constant increased approximately (1.6, 1.6, 1.8, 1.3 and 1.6 folds) in group OI, OI, OIII, BI and BII respectively.

According to $k_{+1}$ values, the rate of reaction in OI is faster than in OII and OIII which may relate to the origin of CA125 tumor marker.

Table (4-6) also shows that reaction rate of the interaction of partially purified CA125 to $^{125}$I-antiCA125 antibody is faster than that in crude CA125 (OI group). The increase in reaction rate is associated with decrease in the reaction time from 240 min. for crude antigen to 180 min and 210 min for BI and BII respectively.

The value of $k_{-1}$ was also determined from the values of $K_a$ (equation 3), which has been estimated at the four temperatures investigated. The results showed that $k_{-1}$ increased with elevation of temperature. The increase in $k_{-1}$ value was about 2.8 folds for all studied groups when the temperature increased from 5 °C to 45 °C.

**Table (4-5): Kinetic parameters for the binding of $^{125}$I-antibodyCA125 antibody with CA125 in ovarian tumor homogenates at different temperatures using second order rate law (All other details are explained in the text).**

| Temperature °C | $k_{+1}$ $(mg^{-1}.ml.min^{-1})$ | $k_{-1}$ $x10^{-3}$ $(min^{-1})$ | $t_{1/2ass}$ (min) | $t_{1/2}$ diss (min) |
|---|---|---|---|---|
| Malignant post-menopausal ovarian tumor | | | | |
| 5 | 0.0070 | 1.288 | 93 | 538 |
| 20 | 0.0083 | 1.900 | 72 | 364 |

| | | | | |
|---|---|---|---|---|
| 37 | 0.0101 | 2.954 | 52 | 235 |
| 45 | 0.0110 | 3.625 | 36 | 191 |
| Malignant pre-menopausal ovarian tumor | | | | |
| 5 | 0.004410 | 1.1970 | 81 | 578 |
| 20 | 0.005580 | 1.8434 | 51 | 375 |
| 37 | 0.006722 | 2.5973 | 27 | 26 |
| 45 | 0.007250 | 3.3989 | 18 | 203 |
| Benign ovarian tumor | | | | |
| 5 | 0.002470 | 1.150 | 123 | 602 |
| 20 | 0.003172 | 1.7622 | 84 | 393 |
| 37 | 0.003925 | 2.5257 | 60 | 274 |
| 45 | 0.004510 | 3.1447 | 27 | 220 |

**Table (4-6): Kinetic parameters for the binding of $^{125}$I-antibodyCA125 antibody to partially purified CA125 BI and BII form of malignant post-menopausal ovarian tumor homogenate at different temperature using second order rate law. (All other details are explained in the text).**

| Temperature °C | $k_{+1}$ (mg$^{-1}$.ml.min$^{-1}$) | $k_{-1}$ x10$^{-3}$ (min$^{-1}$) | $t_{1/2ass}$ (min) | $t_{1/2}$ diss (min) |
|---|---|---|---|---|
| Partially purified (BI) | | | | |
| 5 | 0.04500 | 6.3450 | 48 | 109 |
| 20 | 0.04965 | 9.9300 | 35 | 74 |
| 37 | 0.05620 | 15.288 | 27 | 45 |
| 45 | 0.05937 | 18.550 | 24 | 37 |
| Partially purified (BII) | | | | |
| 5 | 0.01356 | 4.0680 | 72 | 170 |
| 20 | 0.01583 | 5.7560 | 55 | 120 |
| 37 | 0.01860 | 8.2450 | 38 | 84 |
| 45 | 0.02266 | 10.894 | 25 | 63 |

## 4.4.2. The thermodynamic studies of the interaction of $^{125}$I-antiCA125 antibody with CA125.

### 4.4.2.1 Thermodynamic parameters of standard state

Figures (4-5 and 4-6) represents the dependence of equilibrium constant (affinity constant) for binding of [125]I-antiCA125 antibody to crude CA125 of benign, pre-and post-menopausal and partially purified CA125 in ovarian tumor homogenates on the temperature (Van't Hoff plot).

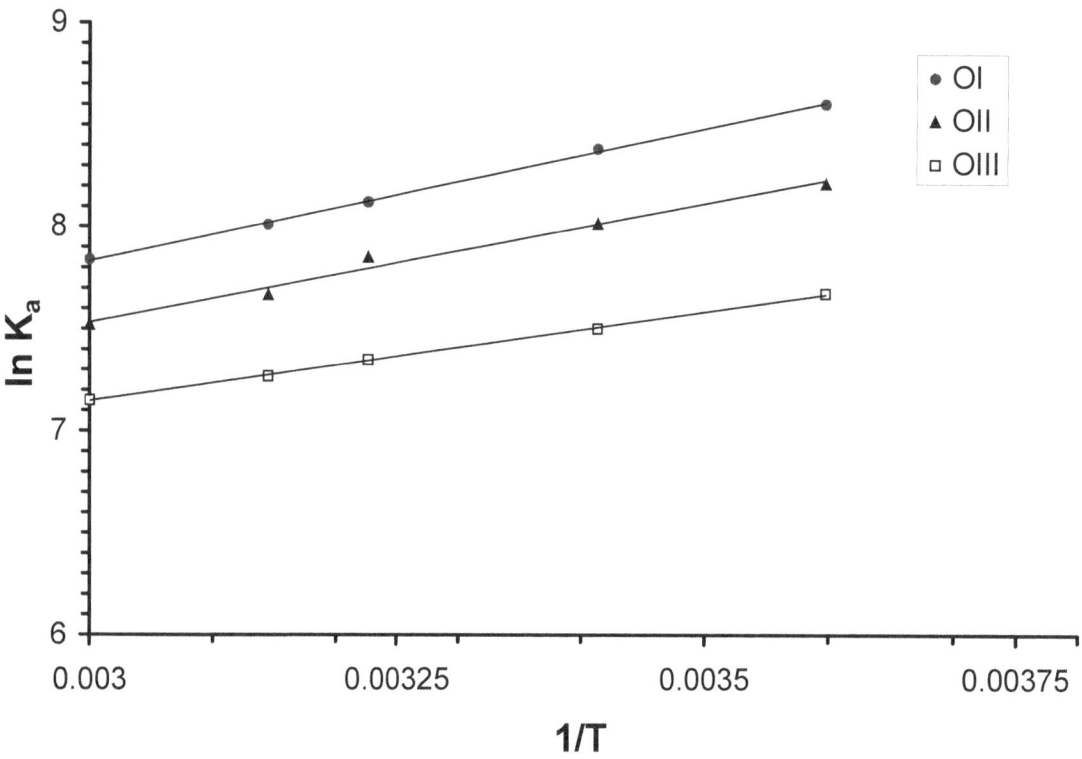

**Figure (4-5): Van't Hoff plot for the binding of CA125 to [125]I-antiCA125 antibody in ovarian tumor homogenates at different temperature for OI, OII and OIII (All other details        are explained in the text)**

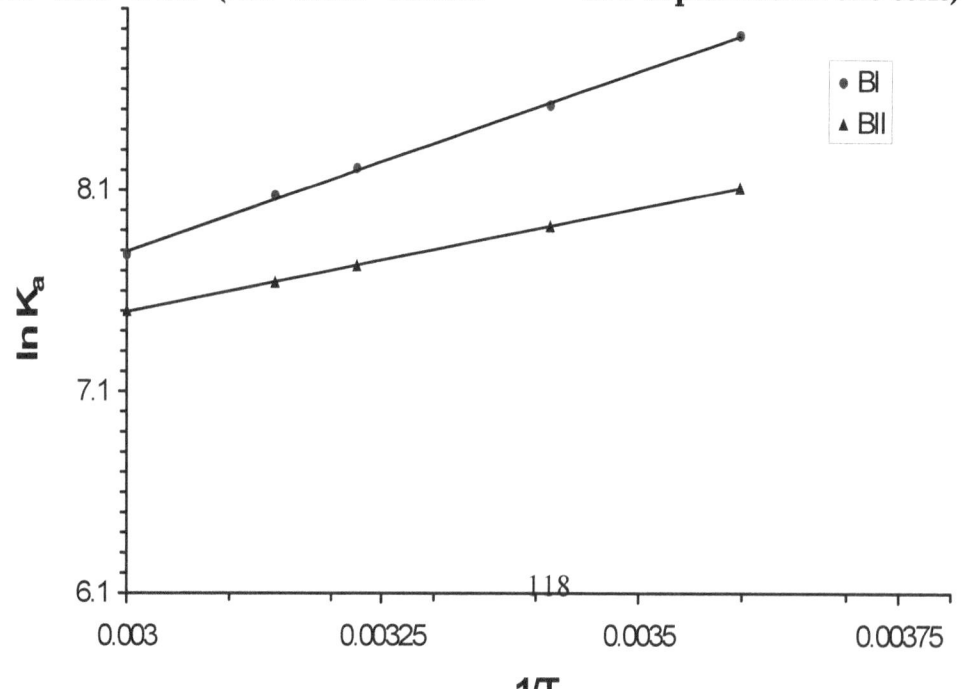

**Figure (4-6): Van't Hoff plot for the binding of partially purified CA125 with $^{125}$I-antiCA125 antibody at different temperatures (All other details are explained in the text).**

Tables (4-7 and 4-8) summarize the calculated thermodynamic parameters for all studied groups (OI, OII, OII, BI and BII) respectively. The results obtained revealed that $\Delta H^0$ in general has small negative value for all groups; their negative sign suggested an exothermic CA125-$^{125}$I-antiCA125 antibody interaction reaction.

The other values of thermodynamic parameters of the standard state such as $\Delta G°$ and $\Delta S°$ at four temperatures, are summarized in the same tables (4-7 and 4-8).

A high value of positive $\Delta S°$ suggests that the reaction spontaneity was entropically driven.

The high negative value of $\Delta G°$ reflects the stability of complex, hence, the high affinity of the reactants. The high negative values of $\Delta G°$ for the binding reaction are controlled by high positively $\Delta S°$ and low negatively $\Delta H°$. [167] so our CA125-$^{125}$I-antiCA125 antibody interaction system is characterized by the contribution of $\Delta S°$ and $\Delta H°$ to the stability of the complex formed.

The high value of positive $\Delta S°$ suggests that the reaction was entropically driven and indicates that the hydrophobic interactions are essentially important in stabilizing the complex [168], while the small negative $\Delta H°$ value may indicate favorable interactions between groups within both CA125 and [125]I-antiCA125 antibody. These include the non covalent interactions which are fundamentally electrostatic in nature such as charge-charge interactions which occur in both CA125 and [125]I-antiCA125 antibody, other types of interactions include charge-dipole, dipole-dipole, charge induced dipole, dipole induced dipole and hydrogen bond. The sum of all types of these interactions can yield sum stabilization to the folded structure of the complex. So the negative value of $\Delta G°$ showed that the overall reaction was energetically favorable in the direction of complex formation.

**Table (4-7): Thermodynamic parameters for standard state of the binding of [125]I-antiCA125 antibody with CA125 in ovarian tumor homogenates at different temperatures.**
**(All other details are explained in the text).**

| Temperature$^0$C | $\Delta H°$(KJ.mol$^{-1}$) | $\Delta G°$(KJ.mol$^{-1}$) | $\Delta S°$(J.mol$^{-1}$ K$^{-1}$) |
|---|---|---|---|
| Malignant post-menopausal ovarian tumor | | | |
| 5 | -10.808 | -19.877 | 32.622 |
| 20 | -10.808 | -20.416 | 32.791 |
| 37 | -10.808 | -20.928 | 32.645 |
| 45 | -10.808 | -21.177 | 32.6069 |
| Malignant pre-menopausal ovarian tumor | | | |
| 5 | -9.342 | -18.978 | 34.616 |

| | | | |
|---|---|---|---|
| 20 | -9.342 | -19.524 | 34.750 |
| 37 | -9.342 | -20.232 | 35.125 |
| 45 | -9.342 | 20.267 | 34.355 |
| **Benign ovarian tumor** | | | |
| 5 | -7.407 | -17.732 | 37.140 |
| 20 | -7.407 | -18.270 | 37.075 |
| 37 | -7.407 | -18.938 | 37.196 |
| 45 | -7.407 | -19.2128 | 37.125 |

Table(4-8): Thermodynamic parameters for standard state of the binding of $^{125}$I-antiCA125 antibody with partially purified CA125 at different temperatures.
(All other details are explained in the text)

| Temperature$^0$C | $\Delta H^{\underline{0}}$(KJ.mol$^{-1}$) | $\Delta G^{\underline{0}}$(KJ.mol$^{-1}$) | $\Delta S^{\underline{0}}$(J.mol$^{-1}$ K$^{-1}$) |
|---|---|---|---|
| **Partially purified BI** | | | |
| 5 | -15.121 | -20.491 | 19.316 |
| 20 | -15.121 | -20.754 | 19.225 |
| 37 | -15.121 | -21.157 | 19.470 |
| 45 | -15.121 | -21.335 | 19.540 |
| **Partially purified BII** | | | |
| 5 | -8.507 | -18.746 | 36.830 |
| 20 | -8.507 | -19.290 | 36.802 |
| 37 | -8.507 | -19.904 | 36.764 |
| 45 | -8.507 | -20.199 | 36.767 |

### 4.4.2.2. Thermodynamic Parameters of Transition State

According to the transition state theory, the interaction of CA125 to CA125 antibody to form the final product 'proceeds through the formation of an activated complex (transition state).

CA125+CA125 antibody $\longrightarrow$ [CA125…CA125antibody] CA125-CA125antibody

An activated complex          Final product
(Transition state)

Arrhenius equation and the kinetic constants have been used to determine thermodynamic parameters of the transition state (Ea, $\Delta H^*$, $\Delta S^*$ and $\Delta G^*$).

Figure (4-7) and (4-8 A&B) shows Arrhenius plots of ln $k_{+1}$ against 1/T values. The slope of the straight line represents the activation energy Ea. Tables (4-9) and (4-10) show the values of thermodynamic parameters of the transition state of all studied groups (OI, OII, OIII, BI and BII). OIII group showed the highest Ea value, which reflects the high energy required to overcome the energy barrier of transition state for the formation of ($^{125}$I-antiCA125 antibody-CA125) complex in comparison to the rest groups. On the other hand the value of activation energy is in accordance with the high positive values of $\Delta G^*$, which indicates that the formation of an activated complex ($^{125}$I-antiCA125 antibody…CA125) is a non spontaneous process and required a lot of energy (equal to Ea) to overcome the transition state energy barrier and giving the final product .Also the positive values of $\Delta G^*$ is mainly attributed to the decrease in entropy of the transition state ($\Delta S^* < 0$) in addition, the positive value of $\Delta H^*$ in all groups shows that the heat content of the activated complex is more than that of isolated species.[169]

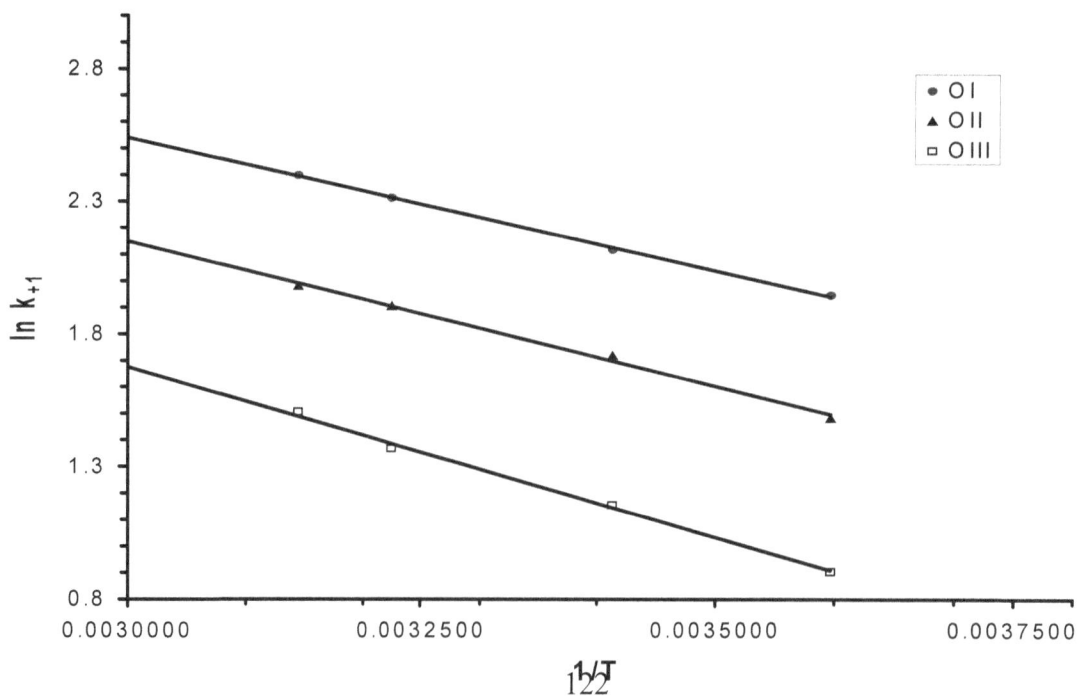

**Figure (4-7): Arrhenius plot of the binding of $^{125}$I-antiCA125 antibody with its CA125 in ovarian tumor homogenates for OI, OII and OIII. (All other details are explained in the text).**

**(A)**

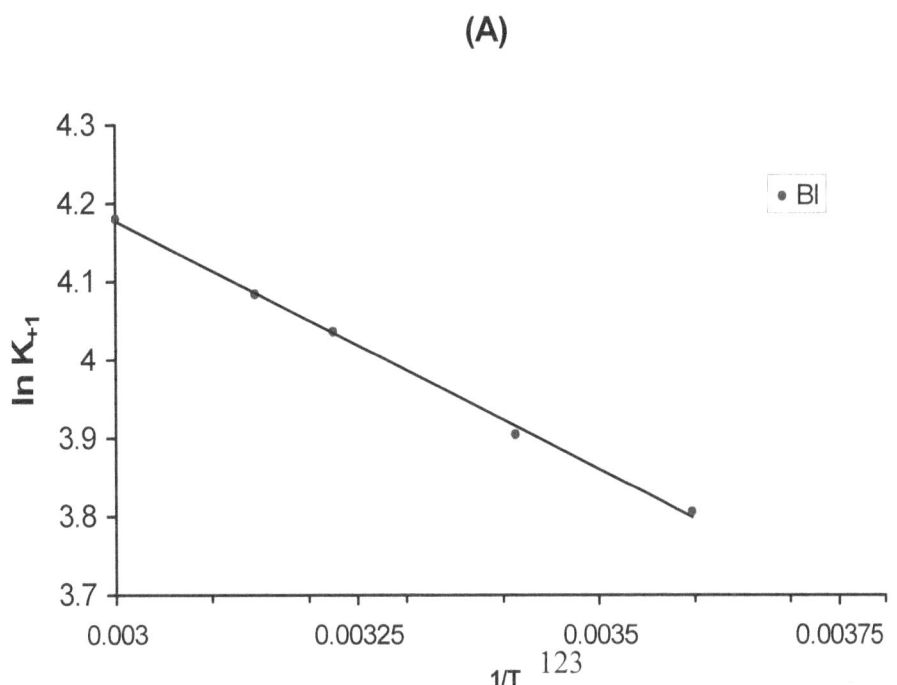

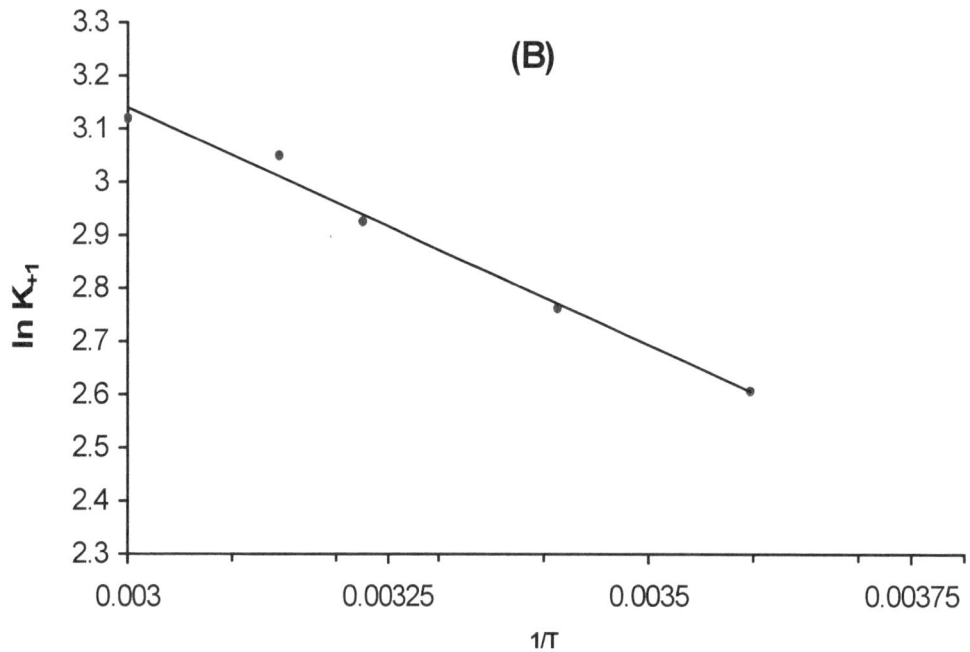

**Figure (4-8): Arrhenius plot for the binding of [125]I-antiCA125 antibody with partially purified CA125 A) BI, B) BII. (All other details are explained in the text).**

Table (4-9):Thermodynamic parameters for the transition state for the binding of $^{125}$I-antiCA125 antibody to CA125 in ovarian tumor homogenates at different temperatures. (All details are explained in the text).

| Temperature °C | Ea(KJ.mol$^{-1}$) | $\Delta H^*$(KJ.mol$^{-1}$) | $\Delta G^*$ (KJ.mol$^{-1}$) | $\Delta S^*$(J.mol$^{-1}$.K$^{-1}$) |
|---|---|---|---|---|
| Malignant post – menopausal ovarian tumor | | | | |
| 5 | 8.132 | 5.820 | 63.418 | -207.187 |
| 20 | 8.132 | 5.696 | 66.561 | -207.730 |
| 37 | 8.132 | 5.554 | 70.053 | -208.061 |
| 45 | 8.132 | 5.489 | 71.713 | -208.251 |
| Malignant pre – menopausal ovarian tumor | | | | |
| 5 | 9.287 | 6.978 | 64.486 | -206.863 |
| 20 | 9.287 | 6.851 | 67.528 | -207.088 |
| 37 | 9.287 | 6.710 | 71.113 | -207.751 |
| 45 | 9.287 | 6.644 | 72.790 | -208.006 |
| Benign | | | | |
| 5 | 10.640 | 8.325 | 65.825 | -206.820 |
| 20 | 10.640 | 8.204 | 68.903 | -207.163 |
| 37 | 10.640 | 8.063 | 73.498 | -207.854 |
| 45 | 10.640 | 7.997 | 75.044 | -207.777 |

Table (4-10): Thermodynamic parameters for the transition state for the binding of $^{125}$I-antiCA125 antibody to partially purified CA125 at different temperatures. (All other details are explained in the text).

| Temperature °C | Ea(KJ.mol$^{-1}$) | $\Delta H^*$(KJ.mol$^{-1}$) | $\Delta G^*$ (KJ.mol$^{-1}$) | $\Delta S^*$(J.mol$^{-1}$.K$^{-1}$) |
|---|---|---|---|---|
| Partially purified BI | | | | |
| 5 | 5.430 | 3.119 | 59.118 | -201.435 |
| 20 | 5.430 | 2.994 | 62.220 | -202.136 |
| 37 | 5.430 | 2.853 | 65.640 | -202.538 |
| 45 | 5.430 | 2.787 | 67.255 | -202.729 |
| Partially purified BII | | | | |
| 5 | 7.156 | 4.845 | 61.890 | -205.197 |
| 20 | 7.156 | 4.720 | 65.004 | -205.747 |
| 37 | 7.156 | 4.579 | 68.482 | -206.138 |
| 45 | 7.156 | 4.513 | 69.802 | -205.311 |

The values of thermodynamic parameters of the binding reaction gave an overall idea about the nature of forces that regulate the formation of complex.

Comparisons of the values of transition state with those of standard state led us to choose a thermodynamic model shown in figure (4-9).

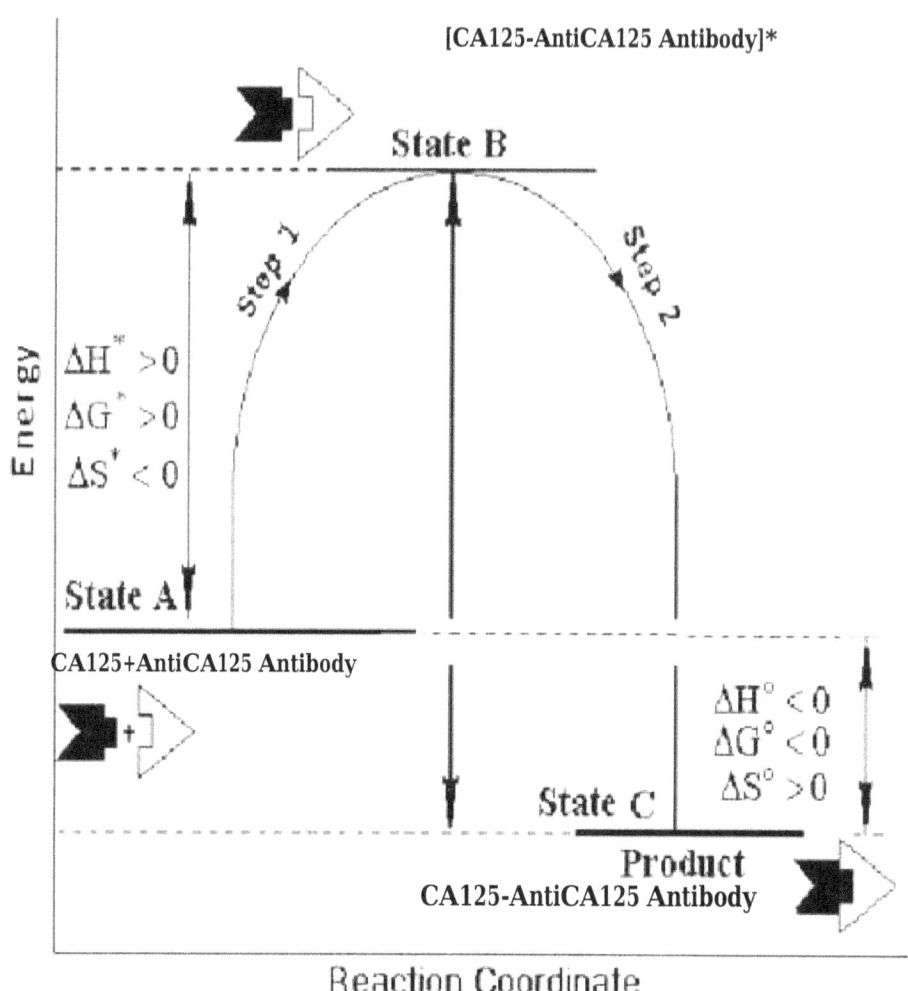

**Figure (4-9): General energy diagram and thermodynamic Model applied to the interaction of $^{125}$I-anti antibody to CA125 of ovarian human homogenates.**

This model proposes that the formation of the ($^{125}$I-antiCA125 antibody/CA125) complex undergoes three thermodynamic state [170]. The thermodynamic state A represents the initial energy level of $^{125}$I-AntiCA125Antibody and its CA125. In the thermodynamic state B, the two species had come together and mutually penetrated their hydration sphere to form a partially immobilized hydrophobically associated species.

Thermodynamic state C represents the fully interacting complex ($^{125}$I-antiCA125 antibody/CA125).

In step 1 of the reaction, the binding of $^{125}$I-AntiCA125Antibody to CA125 was associated with positive $\Delta G^*$ value. This indicates that the initial step of the reaction requires input of energy for the system. The negative entropy change $\Delta S^*$ for this step of reaction reflects the change of the $^{125}$I-antiCA125 antibody/CA125 transition complex to more ordered structure. In step 2, the activated complex participates further interactions, giving the fully interacting complex ($^{125}$I-antiCA125 antibody/CA125).

It is proposed that the formations of a ($^{125}$I-antiCA125 antibody/CA125) complex occurs in these two steps, the first step stabilized by hydrophobic interactions and the second step, as a consequence of hydrophobic interactions, stabilized by short-range interactions such as electrostatic (ionic) interactions, hydrogen bonding, and Vander Waals' interactions which become possible due to the juxtapositioning of appropriate amino acid residues. Although these short-range interactions probably cannot maintain the integrity of the $^{125}$I-antiCA125 antibody/CA125 complex in the absence of the hydrophobic interactions, they are probably responsible for strengthening the stability of the complex. [171]

Hydrophobic interactions contribute to the stability of the complex via high positive excited entropy change ($\Delta S^* > 0$), while electrostatic interactions , hydrogen bonding and Vander Waals' interactions contribute to the stability of the complex   via high negative change in entropy ($\Delta S^* < 0$) [171,172].

The thermodynamic data from this study indicates that the binding of $^{125}$I-antiCA125 antibody to its CA125 are mainly entropically driven and come in agreement with the concept that hydrophobic and short-range interactions have an important role in the binding of $^{125}$I-antiCA125 antibody to its CA125antigen.

# Chapter Five

# Spectroscopic Studies On Isolated ($^{125}$I-anti CA125 antibody/CA125) Complex

Abstract

Spectroscopic studies in the ultraviolet region were carried out to characterize the ($^{125}$I-antiCA125 antibody/CA125) complex of partially purified CA125 antigen (BI and BII) forms which are partially purified from malignant ovarian tumor homogenates.

Gel filtration technique was used to separate $^{125}$I-anti CA125 antibody bound to partially purified CA125 (BI and BII) from unbound (free) $^{125}$I-antiCA125 antibody. Factors affecting the absorption properties of the two types of complexes such as pH, solvent polarity (solvent perturbation technique), spectrophotometeric pH titration and thermal stability in the presence of different concentration of sodium chloride have been studied.

The spectroscopic pH titration curves for complex of partially purified CA125 (BI) and (BII) for histidine residue gave pk$_a$ of 6.5 and 7.2 respectively, while 11.4 and 11.2 for tyrosine residue respectively. Also, it was showed that 60% of histidine and 36.8% of tyrosine residues are located on the surface of complex of (BII) antigen while these residues were 80% and 25% on the surface of complex of (BI).

## 5.1 Introduction

Spectrophotometry is one of the most valuable analytical techniques available to biochemists. Unknown compounds may be identified by their characteristic absorption spectra in the ultraviolet, visible or infrared. The wave length that is absorbed and efficiency of absorption depend on both the structure and environment of the molecule.

The most uses of spectroscopic technique in biochemistry employ UV region spectrum. Proteins absorb U.V light at approximately 280 nm due to the presence of tryptophane, tyrosine and (to a lesser extent) phenylalanine residues within their structure, while the absorption of UV light at (215-230) is due to the presence of the polypeptide chain backbone and histidyl residues [173].

The electronic transitions for these chromophors come from n $\longrightarrow$ $\pi^*$ and $\pi$ $\longrightarrow$ $\pi^*$. Change in the charge and the environment of these chromophors can lead to alteration in the absorption spectrum, and the conformational changes of a protein may also involve environmental change in its chromophoric groups.[174] A large number of environmental factors produce detectable changes in $\lambda_{max}$. Among these factors is the pHs of the solvent which determines the ionization state of ionizable chromophore. Also the solvent polarity affects the chromophore electronic transition where the $\lambda_{max}$ for n- $\pi^*$ transition occurs at shorter wavelength in polar solvents ($H_2O$, Alcohol) than in longer wavelength. The shift may or may not be accompanied by a change in intensity of spectrum. [175, 176] Thus absorbance measurements can give an idea of location of particular amino acid in protein structure.

Although several new Immunochemical techniques were developed to study antigen/antibody interactions [173, 177], U.V spectra remain as one of the most important methods in immunology because it provides a sensitive and quantitative measurements for the study of antibody structure and its specific ligand binding. [178]

Very limited work concerning physical properties of CA125 specially those related to U.V spectroscopy has been done, such as the studies on CA125 in breast cancer by Haider [179] and In colorectal cancer by Al-Jobory [180]. Also the U.V studies on interaction of CA125 with its specific antibody are not wide spread. Hence, the goal of this chapter is to study the spectroscopic behaviour at the region of U.V of the partially purified CA125 and its complex formed with $^{125}$I-antiCA125 antibody at different conditions.

## 5.2 Materials

### 5.2.1 Chemicals

All chemicals and reagents mentioned in section (2.2.1) have been used in the experiments of this chapter.

### 5.2.2. Instruments

All instruments mentioned in section (2.2.2) have been used in the experiments of this chapter.

### 5.2.3. Buffers and Reagents

Buffers and reagents mentioned in section (2.3.4) are used in this chapter. Other additional solutions are indicated in each experiment.

## 5.3 Methods

### 5.3.1. Gel filtration technique for separation of free and bound $^{125}$I- antiCA125 antibody.

#### 5.3.1.1. Preparation of the column.

The dimensions of the column were (1x27cm) chosen according to the equation in section (3.3.1.1).

#### 5.3.1.2. Preparation of the Gel and determination of void volume.

The sepharose CL-6B was used to separate free and bound $^{125}$I-antiCA125 antibody,[157] and was prepared as mentioned in section (3.3.1.3) the void volume was determined and found to be 10.ml.

#### 5.3.1.3. Separation procedure of ($^{125}$I-anti CA125 antibody / CA125) complex.
#### A. Partially purified CA125 (BI) form with $^{125}$I-anti CA125 antibody.

6.      Two hundred micro liter from partially purified CA125(BI) antigen in section (3.3.1.5) (360 $\mu g.ml^{-1}$) was incubated with 80 $\mu l$ of $^{125}$I-anti CA125 antibody (1440 $\mu g.ml^{-1}$), and the reaction was completed to final volume of 600 $\mu l$ with Tris-buffer (0.05 M, pH 6.8). The tubes were incubated for 180 min at 5°C.

7.      At the end of the incubation, the mixture was applied to the surface of a sepharose CL-6B column equilibrated with Tris-buffer 0.05M, pH 6.8. Elution was carried out using the same buffer to separate CA125 bound to $^{125}$I-anti CA125 antibody from unbound antigen with flow rate 12ml.$hr^{-1}$.

8.      The radioactivity of each fraction was counted in gamma counter for 1min.

## Calculations

Radioactivity (c.p.m) for each fraction was plotted against the fraction number.

## B. Partially Purified CA125 (BII) form with $^{125}$I-antiCA125 antibody.

1.      One hundred and eighty micro liter (440 $\mu g.ml^{-1}$) from partially purified CA125 antigen (BII form) in section (3.3.1.5) was incubated with 100 $\mu l$ of $^{125}$I-antiCA125 antibody (1800 $\mu g.ml^{-1}$) and the reaction was completed to final volume of 600 $\mu l$ with Tris-buffer (0.05M pH 7.2). The tubes were incubated for 210 min. at 5°C.

2.      At the end of the incubation, the mixture was applied to the surface of a sepharose CL-6B column equilibrated with Tris-buffer 0.05M pH 7.2. The elution was carried out using the same buffer to separate CA125 bound to $^{125}$I-antiCA125 antibody from unbound antigen with flow rate 12 ml.$hr^{-1}$.

3.      The radioactivity for each fraction was counted in gamma counter for 1 min.

4.      80 $\mu l$ of $^{125}$I-antiCA125 antibody was completed to 600 $\mu l$ with Tris-buffer (0.05M pH 7.2), then this volume was applied to the surface of the column in step 2, and step 2 and 3 were repeated.

## Calculations

Radioactivity (c.p.m) for each fraction was plotted against the fraction number.

### 5.3.2. The U.V Spectrum of ($^{125}$I-anti CA125 antibody/ CA125) Complex, $^{125}$I-antiCA25 antibody and Partially Purified CA125.

#### 5.3.2.1. The U.V Spectrum of ($^{125}$I-anti CA125 antibody / CA125) Complex

The gel filtration profile in section (5.3.1.3 A & B) gave two peaks. The fractions under each peak were pooled and the absorption spectrum was scanned in U.V region against the appropriate blank in reference beam.

#### 5.3.2.2. The UV Spectrum of $^{125}$I- anti CA125 antibody.

Twenty five micro litter of $^{125}$I-anti CA125 antibody was mixed with 475 µl of Tris (0.05M pH 7.2) and placed in 0.5 cm$^3$ cuvette. The absorption spectrum was scanned in the UV region against appropriate blank in the reference beam.

#### 5.3.2.3. The UV Spectrum of partially Purified CA125

Two hundred micro litter from partially purified CA125 antigen (BI) was mixed with 300µl of Tris pH 6.8 and placed in 0.5cm$^3$ cuvette. The absorption spectrum was scanned in UV region against appropriate blank in the reference beam. The same procedure was repeated for partially purified CA125 antigen (BII) using 100 micro liter of (BII) and 400µl Tris-buffer pH7.2 for dilution.

### 5.3.3. Factors Affecting the Absorption Properties of ($^{125}$I-anti CA125antibody / CA125) Complex.

#### 5.3.3.1. The pH Effect on the Complex

#### Reagents

1.      KCl-HCl buffer (pH) was prepared as follows

Solution A: Potassium chloride (0.1M), 0.788 gm was dissolved in a final volume of 100ml deionized distilled water.

Solution B: Hydrochloric acid (0.1 N)

The required pH (3.0) was prepared by mixing 50ml of solution A with an appropriate amount of solution B to obtain the required pH, and then the volume was made up to 100ml with deionized distilled water.

2.        Citrate-phosphate buffer at different pH was prepared as follows.

Solution A: citric acid (0.05M): 0.9605 gm of citric acid dissolved in 100ml deionizer distilled water.

Solution B: Dibasic sodium phosphate (0.1M): 1.4198gm of $Na_2HPO_4$was dissolved in a final volume of 100ml deionized destilled water.

Working buffer pH (4 and 6) was prepared by mixing a volume of solution A with appropriate amount of solution B to obtain the required pH in a final volume of 100ml.

3.        Tris buffer at different pH values was prepared as follows:

Solution A: Tris (hydroxyl methyl amino methan) (0.1M): 1.215gm was dissolved in a final volume of 100ml of deionized distilled water.

        Solution B: Hydrochloric acid (0.1N)

The required pH (7 and 8) was prepared by mixing 50ml of solution A with appropriate amount of solution B to obtain the required pH, and then the volume was made up to 100ml with deionzied distilled water.

4.        Glycine – NaOH buffer was prepared as follows:

Solution A: Glycine (0.1M): 0.7505gm $C_2H_5NO_2$ was dissolved in a final volume of 100ml deionized distilled water.

Solution B: Sodium hydroxide (0.1M):0.4gm NaOH was dissolved in a final volume of 100ml deionized distilled water.

Working buffer pH (9 – 12) was prepared by mixing 50ml of solution A with appropriate amount of solution B to obtain the required pH, then the volume was made up to 100ml with deionized distilled water.

**Procedure**

One hundred micro liter of pooled fractions under the first peak in (Fig 5-1A) and Fig (5-1 B) which represent ($^{125}$I-antiCA125 antibody / CA125) complex of BI and BII respectively were completed to 500 µl with different buffers at different pH values (3, 4, 9 and 12) individually then each sample tube was scanned in UV region against a buffer blank at each pH.

## 5.3.3.2. Effect of Solvent Polarity on UV Spectra of Complex.

The effect of 20% ethanol, and the same amount of ethylene glycol, glycerol and DMSO on the complex was studied.

One hundred micro liter of ($^{125}$I-antiCA125 antibody/CA125) complex of BI and BII [pooled fractions under the first peak in figure (5–1–A) and (5–1–B)] were mixed with 100 µl of either ethanol, ethylene glycol, glycerol and DMSO) separately. The volume was completed to 500 µl with Tris pH 6.8 and pH 7.2 for complex of BI and BII respectively. The absorbance of each sample was scanned immediately in the UV region (200–350 nm) against blank reference contains 20% appropriate solvent.

## 5.3.3.3. Spectrophotometric pH Titration of complex

A Series of complexes from of BI and BII (100µl) were completed to 500µl with buffer at pH ranging from 8 to 12. The absorbance of each sample was measured at 295 nm and the absorbance of $\lambda_{max}$ at each pH value was plotted versus the corresponding pH. Other series of the same complex (100µl) were completed to 500 µl with buffers ranging from 4 to 8. The maximum absorbance of each sample was measured at 211nm and the absorbance of $\lambda_{max}$ at each pH values was plotted against the corresponding pH.

## 5.3.3.4. The Effect of NaCl Concentration on the thermal Stability of the complex by UV spectral studies.

**Reagents**

Buffers used in thermal stability studies of complex of BI and BII were prepared as follows:-

Twenty percent ethylene glycol buffer was prepared by mixing 20ml of ethylene glycol and 80ml of Tris-buffer pH 6.8. NaCl (0.01M) in 20% ethylene glycol buffer was prepared by dissolving 0.05844 gm of NaCl in 100 ml of 20% ethylene glycol buffer, while NaCl (0.1M) in 20% ethylene glycol buffer was prepared by dissolving 0.5844 gm of NaCl in 100ml of 20% ethylene glycol buffer.

20% Ethylene glycol buffer in Tris buffer pH 7.2 was also prepared and used to prepare (0.01M) NaCl and (0.1M) NaCl solution.

## Procedure

One hundred micro litter of complex of BI and BII were completed to final volume 500µl with (0.01M) NaCl in 20% ethylene glycol buffers pH 6.8 and pH 7.2 for complex of BI and BII respectively. Each mixture was placed in 0.5cm cuvette in the sample beam and the buffer at the adjusted pH in the reference beam. The absorption was measured at the wavelength of (292 and 295 nm) at different temperatures 20, 30, 40, 50, 60, 70, 80. The experiment was repeated for each complex with another solution (0.1M NaCl in 20% ethylene glycol) at 292 and 295 nm.

## 5.3.3.5. The Effect of Urea, KCl and Urea - KCl mixture on the spectrum of the complex.

### Reagent

1.  Eight molar urea was prepared by dissolving 24.02 gram of urea in the final volume of 50 ml of Tris pH 6.8

2.  KCl (0.03M) was prepared by dissolving 0.2737 gram of the salt in the final volume of 50 ml of corresponding buffer.

3.	8M urea and 0.03 M KCl solutions were also prepared using Tris pH 7.2 buffers.

## Procedure

One hundred micro liters of complex of BI and BII were pipetted in a set of three tubes. The volume was completed to 500 μl with Tris buffer pH 6.8 for complex of BI and Tris pH 7.2 for complex of BII contains either 0.03M KCl, 8M Urea or mixture 1:1 of both 0.03M KCl and 8M urea respectively. Then each Sample was placed in 0.5cm cuvette in the sample beam and the buffer at the same pH in the presence of the same salt in the reference beam.

The absorption of each sample was scanned immediately in the area of (200 – 350 nm).

## 5.4. Results and Discussion

### 5.4.1. Gel Filtration Technique for Separation of Free and Bound $^{125}$I- anti CA125 antibody

Figures (5-1 A & B) show the results of gel filtration technique to separate $^{125}$I-anti CA125 antibody bound to partially purified CA125 (BI) and (BII) forms respectively. The elution profile in figure (5-1A) revealed two peaks: the first peak with low retention time at fraction number 10 (high molecular weight) represent the complex of partial purified CA125 (BI) with $^{125}$I-anti CA125 antibody bound to $^{125}$I-antiCA125 antibody, while the second peak at fraction number 19 represents the unbound (free) $^{125}$I-anti CA125 antibody. The elution profile in figure (5-1B) also revealed two peaks: the first peak at fraction number 15 represent the complex of partial purified CA125 (BII) bound to $^{125}$I-anti CA125 antibody, while the second peak at fraction number 19 represents the unbound (free) $^{125}$I-anti CA125 antibody. The elution profile for the label antiCA125antibody is shown in Figure (5-2) in which one peak appeared in the same position of second peak of Figure (5-1A&B) which represents the unbound $^{125}$I-antiCA125 antibody. The resultant fractions under the first peak in figures (5-1A) or (5-1B) were collected and pooled.

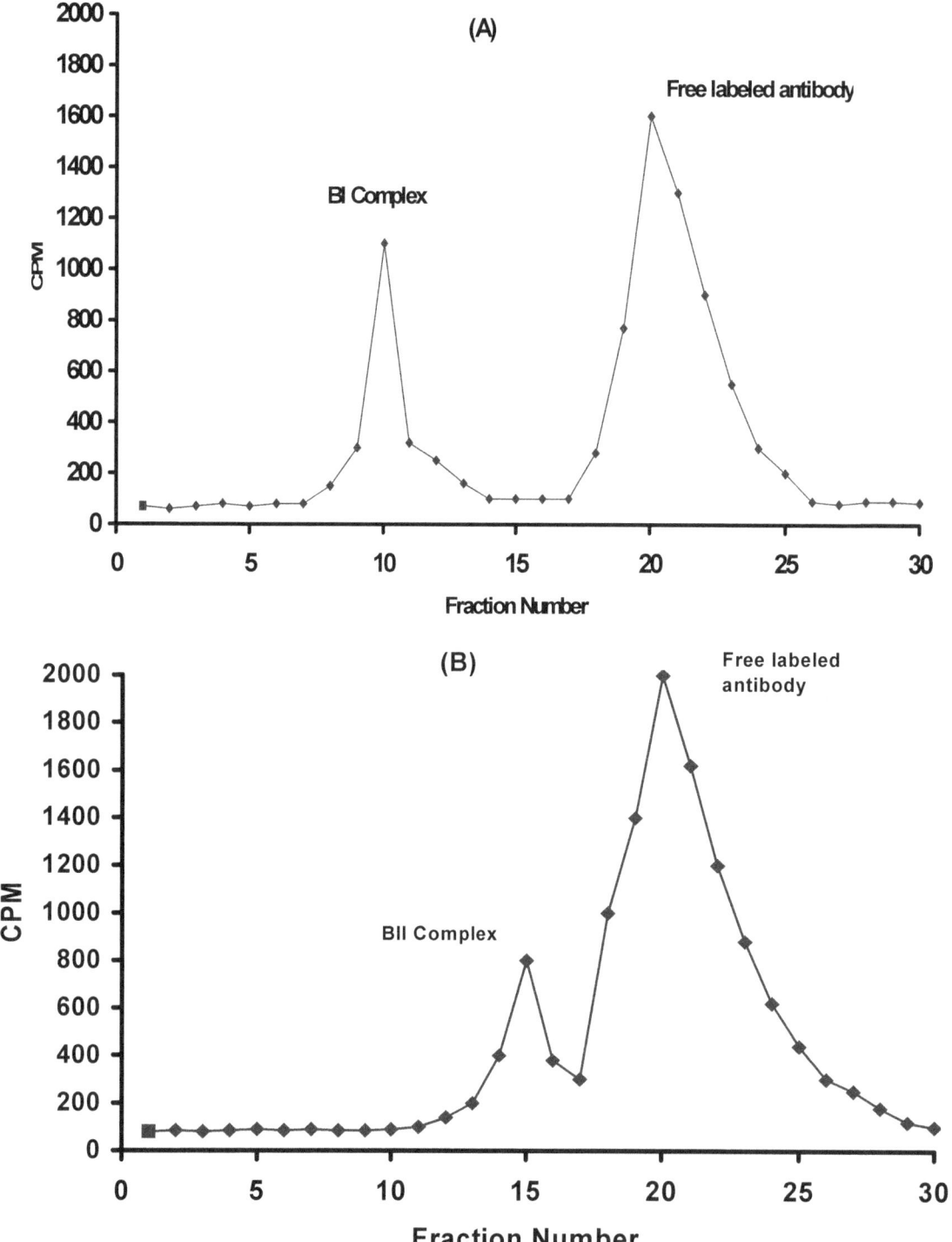

Figure(5-1) The elution profile of the isolated complex ($^{125}$I-antiCA125 antibody/CA125) and free antibody in
(A) Partially purified CA125 antigen ( BI )
(B) Partially purified CA125 antigen (BII ) using sepharose CL-6B gel. (All other details are explained in the text)

**Figure(5-2): The elution profile of free $^{125}$I-antiCA125 antibody/ using sepharose CL-6B gel.(All other details are explained in the text).**

## 5.4.2. The UV Spectra of Partially purified CA125, anti CA125 antibody and ($^{125}$I-antiCA125 antibody/CA125) complex molecules.

The ultraviolet absorption spectra of protein in the regions 250 to 300 nm are contributed from tyrosyl, tryptophan and (to a less extent) phenylalanine residues, but at shorter wavelengths; the contributions come from other groups such as histidyl residues and the peptide bond. The absorbance at lower wave lengths is directly related to the amount of polypeptide material and is usually considerably more sensitive than at 280nm. Absorbance at 215-230nm is useful for monitoring peptides may not contain tryptophan or tyrosine [173].

The UV spectra of partially purified CA125 (BI) and (BII), $^{125}$I-antiCA125 antibody and ($^{125}$I-antiCA125 antibody/CA125) complex were scanned from 200-350nm to determine the absorption spectra and the alteration in the UV spectra as a result of their interaction.

### 5.4.2.1. The UV Spectrum of Partially Purified CA125.

Figure (5-3) and (5-4) shows the UV spectra of partially purified CA125 (BI) and (BII) respectively. The spectrum of (BI) antigen consisted of two

141

peaks; a large one at 220 nm and smaller at 278nm, while the UV spectrum of BI antigen shows two peaks at 212 and 270nm as shown in table 5-1.

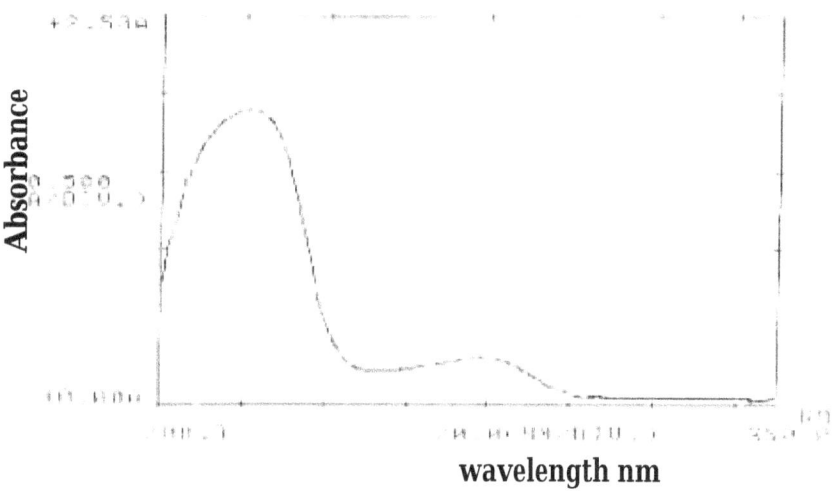

**Figure (5-3): UV Spectrum of partially purified CA125 (BI) (All other details are explained in the text).**

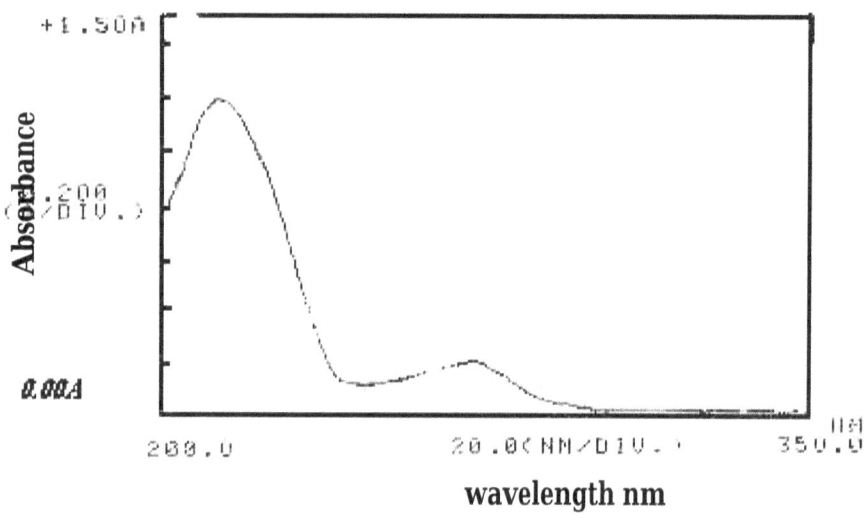

**Figure (5-4): UV Spectrum of partially purified CA125 (BII) (All other details are explained in the text).**

As a result it seemed that each form of CA125 antigen (BI and BII) has a characteristic spectrum and can be identified by its peaks, the first peak at (220 or 212 nm) could be due to the amide group in polypeptide bond of CA125 molecule with contribution of the histidyl residues. While the second peak (at 278 or 270) is assigned to tyrosyl residue.

### 5.4.2.2. The UV spectrum of [125]I- anti CA125 antibody.

The UV spectrum of [125]I-anti CA125 antibody has shown in figure (5-5). The spectrum consisted of two obvious peaks. The first peak at 225nm is assigned to the amide groups in the poly peptide bond with contribution of histidyl residues[173] while the small peak at 278nm is assigned to tyrosyl residue.

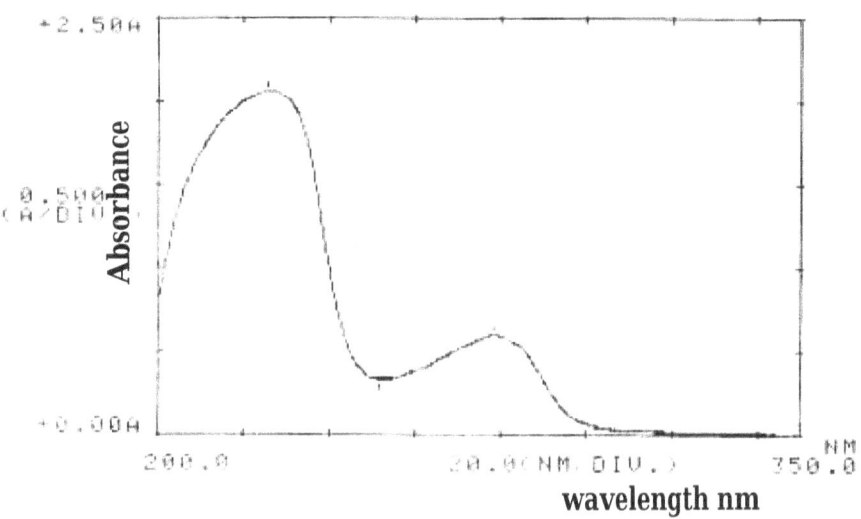

wavelength nm

**Figure (5-5): UV spectrum of [125]I-antiCA125 antibody (All other details are explained in the test).**

### 5.4.2.3. The UV spectrum of ([125]I-anti CA125 antibody / CA125) complex.

Figure (5-6 and 5-7) shows the spectra of partially purified CA125 antigen (BI) and (BII) bound to [125]I-antiCA125 antibody respectively. The spectra of both complexes consisted of one peak at (214 or 208 nm) for the complex of BI and BII of CA125 antigen respectively shown as shown is table (5-1).

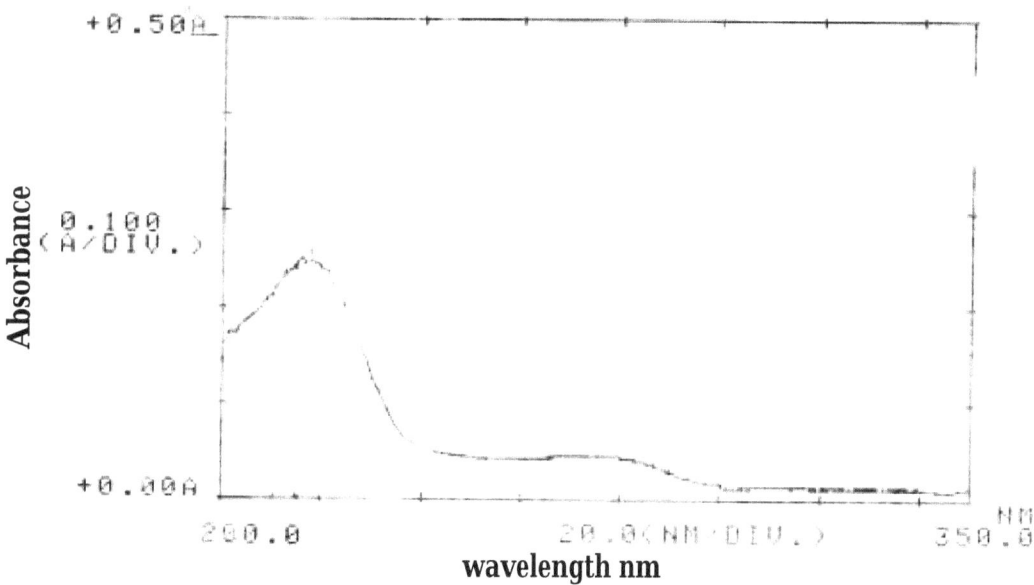

**Figure (5-6): UV spectrum of CA125 (BI)/ [125]I-antiCA125 antibody complex. (All details are explained in the text).**

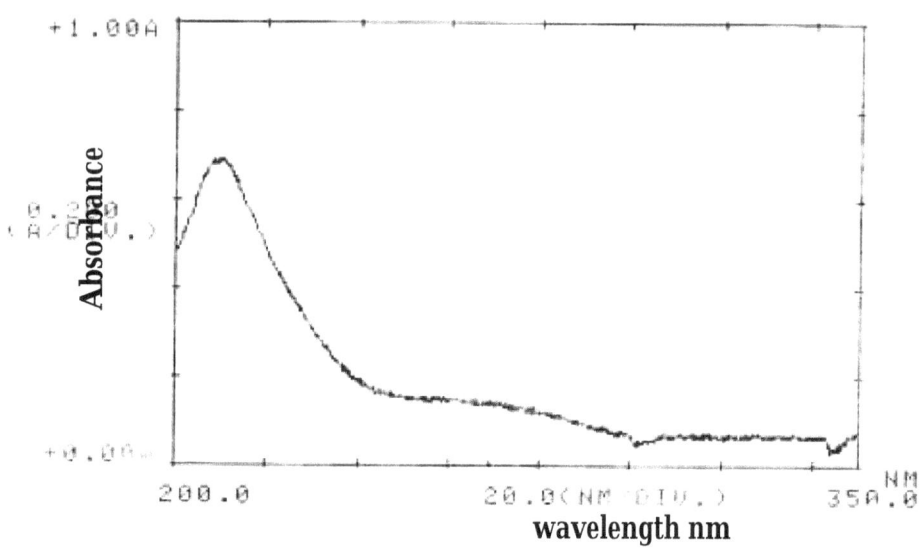

**Figure (5-7): UV spectrum of CA125 (BII)/ [125]I-antiCA125 antibody complex (All details are explained in the text)**

The strong absorption of these peaks (214 or 208 nm) arises from electronic transition in the peptide backbone itself and is therefore sensitive to backbone conformation [181]. There was disappearance of tyrosine peaks in both complexes. These changes are due to fitting of antibody to its antigen to form ([125]I-Anti CA125 Antibody/CA125) complex. This result is in agreement with Seinerman etal observation who found that the surface of protein interactions

144

was polar as well and the complex formation lead to the burial of charged and polar residues. [182]

**Table (5-1): The $\lambda_{max}$ of ($^{125}$I-antiCA125 antibody/CA125) Complex, partially purified CA125 and unbound (free) $^{125}$I-anti CA125 antibody. (ALL other details are explained in the text).**

| No. | Fractions | BI $\lambda_{max}$ (nm) | BII $\lambda_{max}$ (nm) |
|-----|-----------|------------------------|--------------------------|
| 1 | Partially purified CA125 | 220,278 | 212,270 |
| 2 | $^{125}$I-antiCA125 antibody | 225,278 | 225,278 |
| 3 | $^{125}$I-antiCA125 antibody /CA125) complex | 214 | 208 |

## 5.4.3. Factors Affecting the Absorption Properties of ($^{125}$I-antiCA125 antibody/CA125) Complex of BI and BII Antigen.

### 5.4.3.1. The Effect of pH on the complex.

The pH of the solvent determines the ionization state of ionizable chromophore in the protein molecule.[181] Table (5-2) shows the $\lambda_{max}$ values of isolated ($^{125}$I-antiCA125 antibody /CA125) complex of BI and BII at different pH (3, 4, 9 and 12). At an acidic pH 3 both complexes of BI or BII have one maximum wavelength at (205 or 202 nm) respectively compare to the UV spactra of partially purified CA125 antigen (BI) and (BII) or the spectra of $^{125}$I-anti CA125 antibody.

It seems that in acid region there was a blue shift in $\lambda_{max}$ from (214 or 208 nm) in neutral pH  to (205 or 202) in acidic pH for both complexes of (BI) and (BII) respectively. The blue shift is due to the increasing of hydrogen bond formed in the presence of highly positively charged state [183].

When the pH value was increased from neutral to pH 9, there was also only one maximum wavelength found in the spectra of both complexes with the increase in $\lambda_{max}$ value from ( 214 or 208) to ( 223 or 220) for complex of BI and BII

145

respectively. At pH 12 two $\lambda_{max}$ were obtained for each complex. $\lambda_{max1}$ and $\lambda_{max2}$ of complex of (BI) were at (225, 284nm) respectively whereas $\lambda_{max1}$ and $\lambda_{max2}$ of complex of BII were (223, 282nm) respectively.

These results indicate that tyrosine residue in both complexes molecules at neutral pH are located in a way that a small part of it is on the surface of the protein molecule while a large part of these residues is buried but at high pH = 12 the protein becomes denatured (unfolded) and the internal tyrosine has become exposed to the solvent and absorb light. A red shift observed in absorption of these tyrosine residues is certainly related to the ionization of side chain of the tyrosine and this led to the availability of the lone pair on the oxygen atom to happen easier and at lower energy level (red shift).

**Table (5-2): The effect of different pH on $\lambda_{max}$ value of ($^{125}$I-anti CA125 antibody/CA125) complex. (All details are explained in the text).**

| pH | $\lambda_{max}$ ($^{125}$I-antiCA125 antibody / CA125) (BI complex) | ($^{125}$I-antiCA125 antibody / CA125) (BII complex) |
|---|---|---|
| 3 | 205 | 202 |
| 4 | 212 | 207 |
| 7 | 214 | 208 |
| 9 | 224 | 221 |
| 12 | 225, 284 | 223, 282 |

### 5.4.3.2. The Effect of solvent polarity on the UV spectrum of the complex

The determination of whether an amino acid is internal or external by measuring the spectra of a protein in polar and non- polar solvent is called the solvent perturbation method. In fact, proteins are rarely studied in completely non- polar solvents. However, significant solvent effects can be induced by the use of a mixture of water and substance of a reduced polarity such as ethanol, ethylene glycol, glycerol and dimethylsulfoxide [144].

Several spectral changes were obtained in the presence of these perturbants, like the alteration of the peak position and intensities of protein spectrum, and the appearance of new chromophores on the surface of protein molecule. These chromophores were embedded in an interior region of the protein in the absence of the solvent. One of the main assumptions of the solvent perturbation technique is that solvents alter the peak position and intensities by altering the energy and probability of electronic transitions.[184-185] In reality the preferential solvation is caused not only by the perturbant interaction with chromophor itself, but also with the group adjacent to the chromophor in the protein.

Table (5-3) shows the effect of different solvents on the ($^{125}$I-antiCA125 antibody/CA125) complex of (BI) and (BII) form. It was found that ($\lambda_{max}$ 214 or 208 nm) shown in previous experiments for complex of BI or BII respectively, was shifted to longer wavelength (red sheft) in the presence of ethanol, ethylene glycol and glycerol at a concentration of (20%). These shifts are attributed to the amide group in polypeptide bond with contribution of histidyl residues inter molecular hydrogen bonding between amide group of polypeptide bond in the complex molecule and the solvent may cause these shifts the intermolecular hydrogen bonding increase as the concentration of the solution increase and additional bands start to appear at longer or shorter wavelength [186]. The contribution of histidyl residues in the observed spectra $\lambda_{max1}$ is difficult to detect because of the overlapping of its absorbance with that of peptide bond.[176] Table (5-3) shows that the spectrum of both protein complexes are is sensitive to change in the polarity of the solvent which may indicate that high percent of histidine residues is located on the surface of the protein molecule.

In the presence of DMSO (20%) there was an increase in $\lambda_{max1}$ from (214 or 208) to 242 nm for both complexes and new $\lambda_{max}$ at (284, 282 nm) appeared which belongs to complexes of BI and BII respectively. The new peaks are related to tyrosyl residues. The appearance of $\lambda_{max}$ of tyrosine residue is related to the perturbing solvent which makes the possibility for the presence of this residue to the surface of the protein structure.

**Table 5-3: The effect of solvent polarity on $\lambda_{max}$ of ($^{125}$I-antiCA125 antibody / CA125) complex**
**(All other details are explained in the text)**

| Solvent 20% of | ($^{125}$I-antiCA125 antibody/CA125) (BI) | | ($^{125}$I-antiCA125 antibody/CA125) (BII) | |
|---|---|---|---|---|
| | $\lambda\,max_1$ | $\lambda\,max_2$ | $\lambda\,max_1$ | $\lambda\,max_2$ |
| Ethanol | 222 | - | 219 | - |
| Ethylen glycol | 217 | - | 215 | - |
| Glycerol | 223 | - | 221 | - |
| DMSO | 242 | 284 | 242 | 282 |

### 5.4.3.3. Spectrophotometric pH Titration of the Complex.

Spectrophotometric pH titration is the following of the changes in absorbance of the chromofor with increasing pH [144]. Many studies of protein structure require the determination of $Pk_a$ values for protein dissociation from ionizable amino acid side chains, because these values give an indication of the location of the amino acid in the protein. This can often be done spectrophotometricaly because dissociation often changes the spectrum of one of the chromophores, the observation of tyrosine dissociation was performed by measuring the absorption at 295 nm ($\lambda_{max}$ for the ionized form of tyrosine), and the observation of histidine dissociation was carried out by measuring the absorption at 211 nm

The titration curve of ($^{125}$I-antiCA125 antibody/CA125) complex of (BI) and (BII) for both tyrosyl and histidyl residues are illustrated in figure (5-8 A&B) respectively. Figure (5-8A) shows that the $Pk_a$ for tyrosine is 11.4 for ($^{125}$I-Anti CA125 Antibody / CA125) complex of (BI), while the $Pk_a$ for tyrosine is 11.2 for ($^{125}$I-antiCA125 antibody / CA125) complex of (BII) form. From the same curves it could be concluded that about (36.8%) and ( 25%) of tyrosyl residues are located on the surface of protein complex of BI and BII respectively. The other residues are buried interiorly in a polar environment of protein complex of both forms of CA125 antigen. A large arise in the absorbance was observed at high pH because protein complexes become denatured.

Figure (5-8B) shows that $Pk_a$ values of histidine residues in complex from BI and BII antigen are (6.5) and (7.2) respectively. Also from these curves it was found that (80%) histidyl residues are located on the surface of the complex of BI antigen, while (60%) located on the surface of the complex of BII antigen. the other histidine residues are buried interior the protein complex of both antigens.

These results are in agreement with those found in solvent perturbation experiment in section (5.4.3.2.) for the percent of histidine residue on the surface of protein complex molecule.

**(A)**

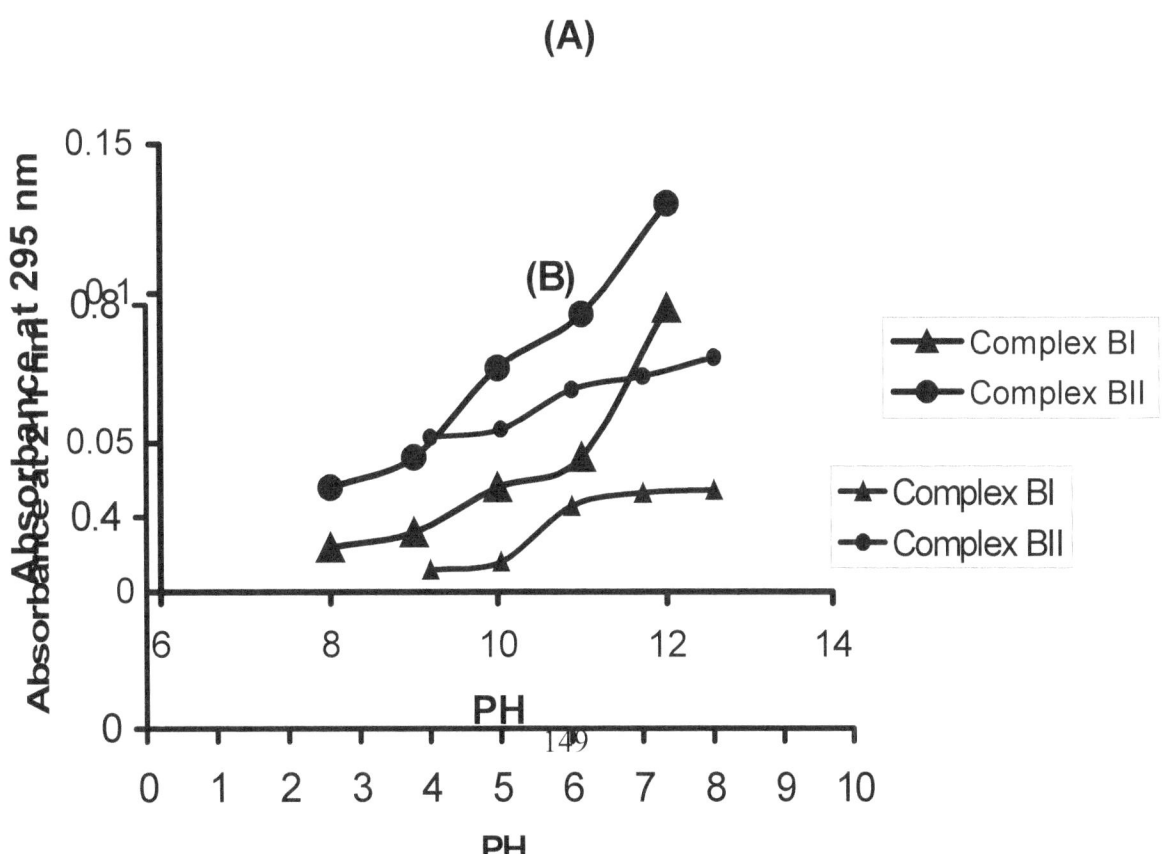

**Figure (5-8): Spectrophotometric pH titration of ($^{125}$I-antiCA125 antibody/CA125) complex from BI and BII antigen.**
**A=for tyrosine, (B) for Histidine.**
**(All other details are explained in the text).**

## 5.4.3.4. The effect of NaCl concentration on the thermal stability of the complex by UV spectral studies.

The effect of the different concentration of NaCl on the thermal stability of ($^{125}$I-antiCA125 antibody/CA125) complex of BI and BII of partially purified CA125 antigen was examined in this experiment. The values of absorbance at $\lambda_{max}$ (292, 295 nm) for tryptophyl and tyrosyl residues respectively, in two different concentrations of NaCl 0.01 M and 0.1 M in 20% ethylene glycol buffer are shown in figure (5-9A&B) and (5-10A&B).

As shown in figure (5-9A&B) the internal tryptophane and tyrosine are completely exposed to the solvent at 60°C in the presence of 0.01 M NaCl in both complexes of BI and BII. The increment in the absorbance of both tryptophyl and tyrosine residues with increasing temperature could be due to those buried chromophores becomes exposed to the solvent during thermal denaturation [187].

150

Figure (5-10 A&B) shows that in presence of 0.1 M NaCl, the absorbance of both tryptophan and tyrosine reach higher value at 70°C for both complexes of (BI) and (BI), therefore higher concentration of NaCl causes more stabilization for protein complex .

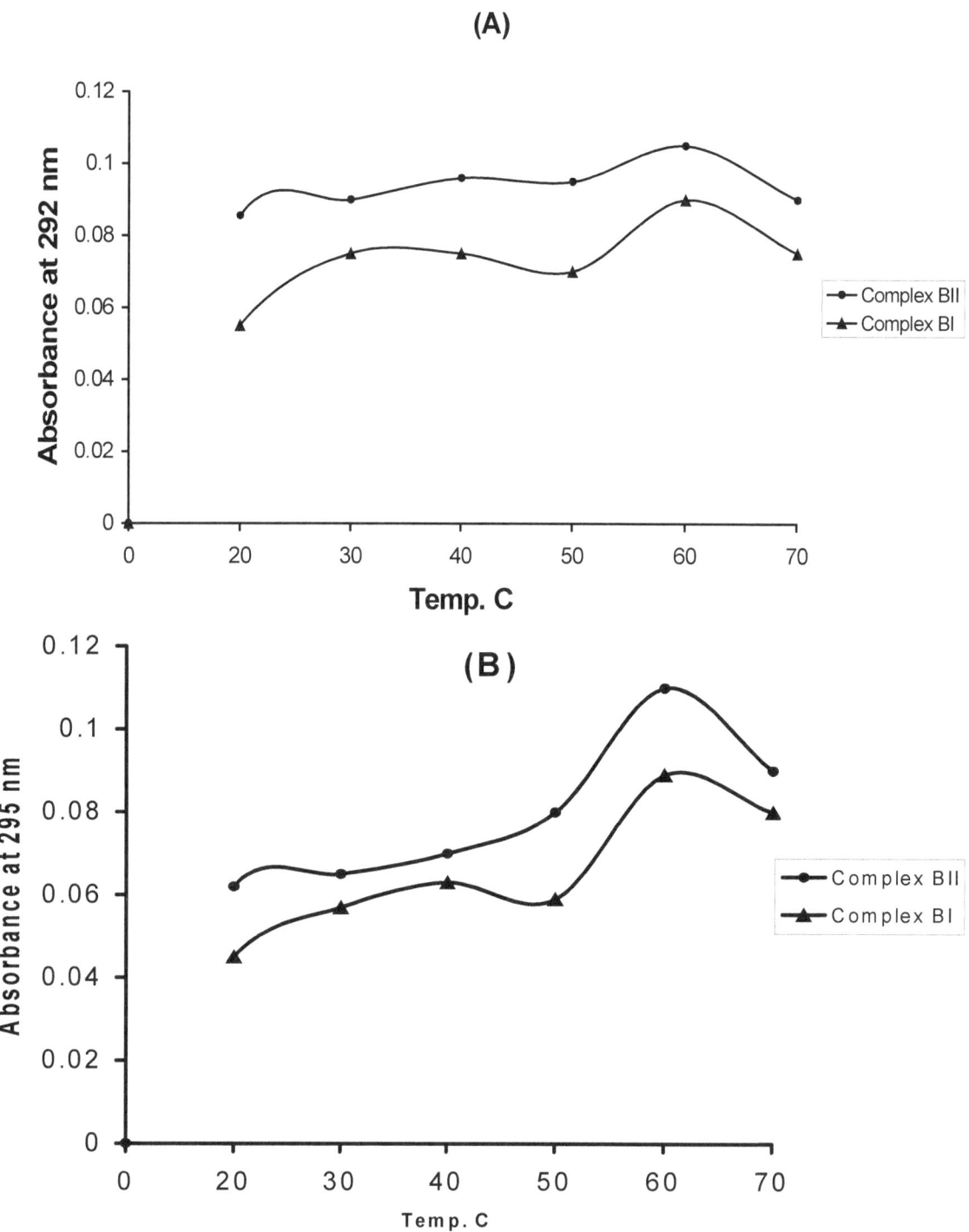

Figure (5-9): Thermal stability curve for complex of BI and BII
(A) at $\lambda_{max}$ 292 in presence of 0.01M NaCl,
(B) at $\lambda_{max}$ 295 in the presence of 0.01M NaCl.
(All other details are explained in the text).

The decrease of the absorbance in presence of 0.1M NaCl as compared with that in 0.01M NaCl could be due to salt concentration. Each protein in solution containing salts will collect around it a counter ion atmosphere enriched in oppositely charged small ion, (chloride ion and sodium ion), and such a cloud of ions will tend to screen the protein, the larger the concentration of small ion present, the more effective this electrostatic screening will be, and decrement in the absorption intensity will be observed [181].

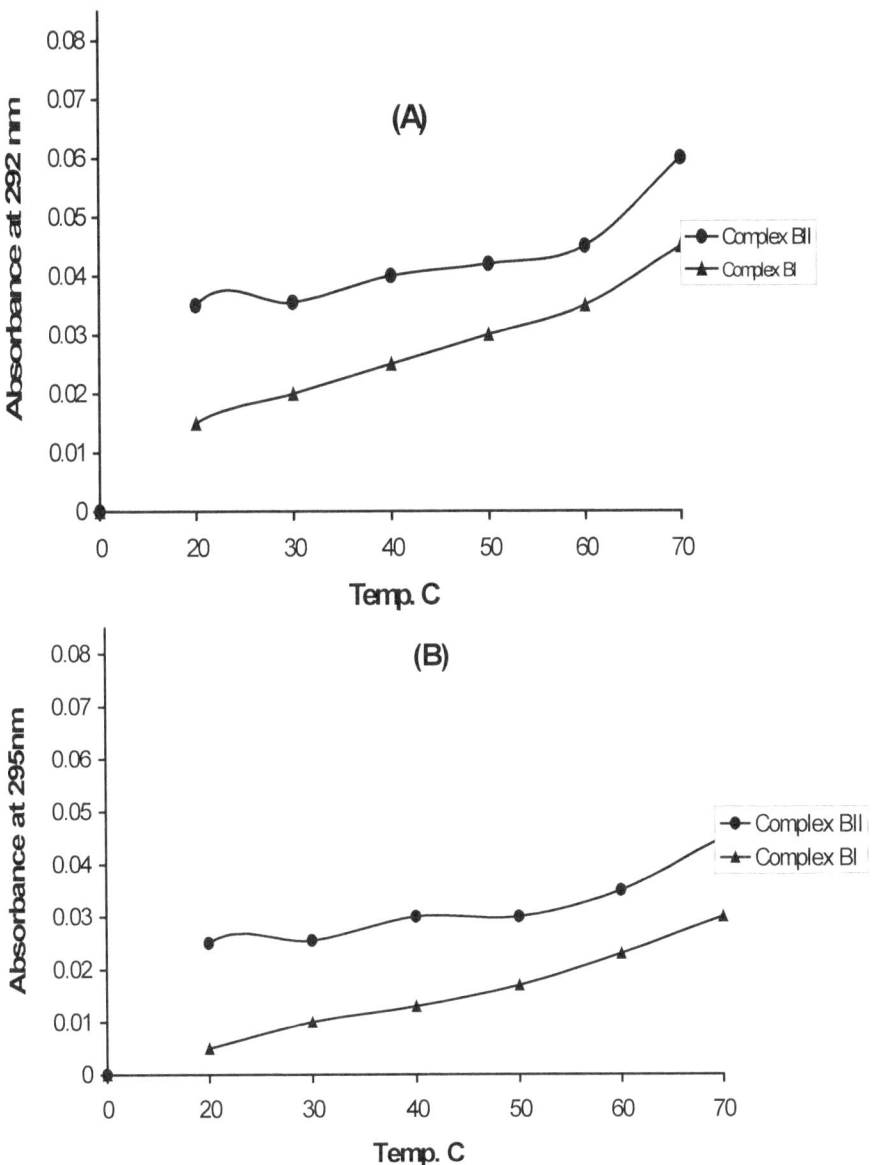

**Figure (5-10 A&B): Thermal stability curve for complex of BI and BII:**
(A) at $\lambda_{max}$ 292 in presence of 0.1M NaCl,
(B) at $\lambda_{max}$ 295 in presence of 0.1M NaCl.
(All other details are explained in the text).

*5.4.3.5. Effect of Urea, KCl and (Urea- KCl) Mixture on the Spectrum of the Complex*

153

Table (5-4) shows the effect of 8M urea, 0.03M KCl and a mixture of 1:1 of 8 M urea and 0.03 M KCl on $\lambda_{max}$ of the complexes of both forms of partially purified CA125 antigen. Comparing the values of $\lambda_{max}$ of these molecules obtained in the absence of urea or KCl (table 5-1) with those obtained in the presence of 8M urea in table (5-4), it seems that there was a significant red shift in $\lambda_{max1}$ of poly peptide bond from (215 or 208nm) to $\lambda_{max}$ (223 or 221 nm) for complex from BI and BII antigen respectively. While there is no $\lambda_{max2}$ peak assign for aromatic amino acid, i.e. tyrosine residue in both complexes. These results indicate that the molecule solvated with urea (dipole –dipole interaction) and produce a red - shift and new chromofore come to the surface .The red shift is due to the intermolecular hydrogen bonding between the oxygen of the amide group and the solvent [176]. When 0.03M KCl was used, there was a slight blue shift (3-1nm) in $\lambda_{max1}$ of polypeptide bond in the complex of BI and BII antigen respectively such blue shift can arise by introducing positive ($K^+$) or negative ($Cl^-$) charges near the chromophore (the amide group, which might interact with $\pi$ – electron system of the amide group [174].

When a mixture of 1: 1 of 8M urea and 0.03 M KCl was used, there was a significant red shift in $\lambda_{max}$, (215 or 208) to $\lambda_{max}$ (221 or 218nm) in complex of BI and BII respectively. The same shift in $\lambda_{max}$ was observed when 8M urea was used alone with each complex. This means that the shift caused by mixture due to the effect of urea but not to 0.03M KCl. Solvent perturbation or denaturation of protein produces many changes in absorption near 230 nm and 280 nm, they are usually about ten times greater near 230nm than at 280nm when the native and the denatured protein are compared. [2] Some of these changes in absorption may be produced by the change in n – $\pi^*$ absorption of poly peptide bond in protein either because of a change in their geometrical arrangement or because of environment changes [181].

**Table (5-4): The effect of 8M Urea, 0.03M KCl and mixture (Urea-KCl) on the $\lambda_{max}$ of complex UV spectrum (All other details are explained in the text).**

| Solvent | $\lambda_{max}$ (nm) | |
|---|---|---|
| | complex of BI | complex of BII |
| Urea 8M | 223 | 221 |
| KCl 0.03M | 211 | 207 |
| Urea-KCl Mixture 1:1 | 221 | 218 |

## Conclusions

Out of the results obtained in this work we can conclude the following:

1.     CA125 is more specific tumor marker for ovarian malignancy than CEA. A higher ratio of CA125 to CEA could be used to differentiate ovarian from non ovarian malignant diseases when both sera contain increased CA125 concentration.

2.     The modified protocol for the assay of CA125 is suitable for the assessment of CA125 in the tissues of ovarian tumor homogenates.

3.     Partial purification of CA125 showed two forms (BI) with 670 kDa molecular weight and (BII) with 100 kDa molecular weight.

4.     The kinetic studies of $^{125}$I-antiCA125 antibody with CA125 of benign and malignant ovarian tumor tissues and with partially purified CA125 of malignant post-menopausal ovarian tumor tissues revealed that the binding reactions are time and temperature dependent. The binding data fits the second order reaction kinetics at (5, 20, 37 and 45 °C) in all cases.

5.     Thermodynamic studies for the binding reaction of $^{125}$I-antiCA125 antibody with CA125 in benign and malignant ovarian tumor tissues and with partially purified CA125 of malignant post-menopausal ovarian tumor tissues, revealed that the binding reactions were exothermic ($\Delta H° < 0$) and spontaneous ($\Delta G° < 0$) and the binding reactions were entropically and enthalipically driven ($\Delta S° > 0$) and ($\Delta H° < 0$).

6.     The spectroscopic studies revealed that partially purified forms of CA125 and their complexes with $^{125}$I-antiCA125 antibody have characteristic spectra and give an idea of the location of particular amino acids in ($^{125}$I-antiCA125 antibody/CA125) complex molecules.

Future work

According to the results obtained in this work, the following works are suggested for the future:

1.        Manufacturing of IRMA kit, using the optimum conditions concluded out of this study's results, to determine CA125 in ovarian tumor tissues.

2.        Application of the modified IRMA method for the assessment of CA125 in other tissues such as lung and pancreas.

3.        Assessment of new markers adjuvant to CA125 to be used for early diagnosis and follow up of the patients with ovarian cancer.

4.        Additional works are needed for farther purification of partially purified CA125 from ovarian tissues using high performance liquid chromatography (HPLC) based on affinity chromatography or ion-exchange technique. Then using the purified CA125 antigen to produce specific antiCA125 antibody and $^{125}$I-CA125 tracer.

5.        Development of radioreceptor-assay protocol to determine the receptors of CA125 in ovarian tissues, then molecular characterization of these receptors.

References

1.	White A, Handler PH, and Smith E. **Principles of Biochemistry**; 5th Ed; Mc Gray-Hill. 1972; pp.812.

2.	Ablev GI. **Alpha-feto protein in ontogenesis and its association with malignant tumors**. Adv Cancer Res 1971; 14:292.

3.	Gold P, and Freeman SO .**Demonstration of tumor –specific antigen in human** colonic carcinoma by immunological tolerance absorption technique J Exp Med 1965; 121:439-450.

4.	Kohler G, and Milstein. **Continuous cultures of fused cells secreting antibody of predefined specificity Nature** 1975; 257(5517):495-497.

5.	Kreuzer H and Massay A. **Recombinant DNA and Biotechnology;** ASM press comp, USA, 1996, chapter 24, pp. 190-198.

6.	Hayes DF, Bast R, Desch CE, Fritsche H, Kemeny NE, et al . **A tumor marker utility grading system (TMUGS). A framework to evaluate clinical utility of tumor markers**. Nati Cancer Inst 1996; 88:1456.

7.	Bates S E and Longo D L .**Use of serum tumor markers in cancer diagnosis and management.** Semin Oncol. 1987(2)102.

8.	Stearns V, Yamauchi H, and Hayes DF. **Circulating tumor markers in breast cancer: Accepted utility and novel prospects.** Breast Cancer Research and treatment. 1998; 52:239-259.

9.	Costa J, and Cordon CC. **Cancer Diagnosis: Molecular Pathology; in: cancer principles and practice of oncology**; 6th ed.; De Vita V.T., Hellman S., Rosenberg Eds. Lippincott Williams and Wilkins. 2001; pp.641-657.

10.	Anderson SC, and Cockayne S. **Clinical Chemistry; concepts and applications**; An HB. Jint .ed; Philadelphia; W.B. Saunders company; 1993; chapter 16, pp.322-330.

11.	Moossa AR, Schempff SC, and Robson MC. **Comprehensive Textbook** of Oncology; 2nd ed.; Eds. Baltimopre, Williams & Wilkins, 1991; pp.225-238.

12.      Haber D and Fearon E. **The promise of cancer genetics**. Lancet, 1998; 351:SII1.

13.      Orr-weaver TL and Weinberg RA. **A checkpoint on the road to cancer**. Nature 1998; 392:223-224.

14.      Brown JM, and Wouters BG. **Apoptosis, p53 and tumor cell sensitivity to anticancer antigen. Cancer Res** 1999; 59: 1391-1399.

15.      Kastan MB, Onyekwere O, Sidransky D, Vogelstein B, and Graig RW. **Participation of p53 in the cellular response to DNA damage**. Cancer Res.; 1991; 51: 6304.

16.      Wu J. **Diagnosis and management of cancer using serologic tumor markers. In; Clinical Diagnosis and Management by Laboratory Methods**; 20th ed.; J.B. Henry Ed. W.B. Saunders Company; 2001; pp.1028-1042.

17.      David W. **Immunoassay Hand Book**; 2nd ed.; U.K. Nature Publishing Group, 2001; chapter 63, pp.635-662.

18.      Zurawski VR, Broderick SF, Pickens P, et al. **CA125 levels in group of non hosptalised women: relevance for the early detection of ovarian cancer**. Obstet Gynecol 1987; 69:606-611.

19.      **European     Group     of     Tumor     Markers     (EGTM)**. http://www.med.uni.muenchen.de/egtm/detail/4.htm .

20.      Thomas j, Nowak A, and Gordon H. **Essential of Pathophysiology**; 2nd ed. WcB Mc Graw-Hill; 1999; pp. 501-504.

21.      Eric V Mackay, Norman A Beischer, Lloyd W Cox, and Carl wood. **Illustrated text book of Gynecology**; W.B. Saunders Company 1983, p. 282.

22.      Cotran RC, Kumor V, and Collins T. **Robbins Pathologic Basis of Disease**; 6th ed.; WB Saunders Company; 1999; pp.1068-1069.

23.      Carol Mattson Porth; **Pathophysiology**; 4th ed.; J. B. Lippincott company Philadelphia; 1994, p.749.

24.      Fuys Woodruff J D; **Pathology: Practical Gynecologic Oncology**; 2nd

ed.; Williams and Wilkins Company; 1995; p.1079.

25.     Kristensen GB and Trope C. **Epithelial ovarian carcinoma.** Lancet 1997; 349:113-117.

26.      Scott JR, Disaia PJ, Hammond CB, and Spellacy WN. **Danforth's Obstetrics and Gynecology**; 8th ed.,; Lippincott Williams & Willkins ; 1999; pp 678-695.

27.      Rosa J. **Ackerman's Surgical Pathology**; 8th ed. Atimes Mirror Company; 1996; p1474.

28.      Caroline Van Haaften and Cinda M Boyer. **Epithelial ovarian tumors in the reproductive age group: Age is not an independent prognostic factore.** Cancer 1996; 77: 1131.

29.      Karlan BY and Platt LD. **The current status of ultrasound and color Doppler imaging in screening for ovarian cancer** .Gynecol Oncol.   1994; 55:528

30.      Charles R Whitefield; **Dewhurst's text Book of Obstetrics and Gynecology**; 5th ed.; Black well science; 1995; PP.759-774.

31.      Tiernery JR, Mcphee SJ, and Oapadakis Mia.; **Current Medical Diagnosis & Treatment**; 38th ed.; Appelton & Lange; 1999; P. 718.

32.      Parkin DM, Whelan SL, Ferlay J, Raymond L, and Young J. **Cancer incidence in five continents**; vol VII Lyon: International Agency for Research on cancer 1997 (IARC scientific Publication No. 143).

33.      Greenlee RT, Hill-Harmon MB, Murray T, and Thun M. **Cancer statistics.**        CA Cancer J. Clin 2001; 51:15.

34.     **Iraqi cancer registry** 1995-1997.

35.      Piver MS, Baker TR, Piedmote M, and Sandecki AM .**Epidemiology and etiology of ovarian cancer.** Seminars in Oncology 1991; 18 (3): 177-185.

36.      Kerlikowske K, Brown JS, and Grady DG. **Should women with familial ovarian cancer undergo prophylactic oophorectomy?** Obstet Gynecol; 1992; 80:700.

37.     Serova O, Montagna M, Torchard D, et al. **A high incidence of BRAC1 mutations in 20 breast-ovarian cancer families.** Am J. Hum. Genet; 1996; 58:42

38.     Wooster R, Bignell G, Lancaster J et al. **Identification of breast cancer susceptibility gene BRAC2.** Nature 1995; 378: 789.

39.     Gillis CR, Hole DJ, still RM, Davis J, and Kaye SB. **Medical audit, cancer registration and survival in ovarian cancar.** Lancet 991; 337: 611-612.

40.     Yancik R, Ries LG, and Yates JW. **An analysis of surveillance, epidemiology, and end results program data.** Is J Obstet Gynecol 1966; 154:636-47?

41.     Whittermore AS, Harris R, and Itnyre J. **Characteristics relating to ovarian cancer risk: Collaborative analysis of twelveU.S.case-controlstudies .II Invasive epithelial ovarian cancers in white women.** Am. J. Epidemiol 1992; 136: 1184.

42.     Gross TP and Schlesselman JJ. **The estimated effect of oral contraceptive use on the cumulative risk of epithelial ovarian cancer.** Obstet Gynecl 1994; 83:419.

43.     Risch HA, Marrett LD, and Howe GR. **Parity, contraception, infertility and risk of epithelial ovarian cancer. Am.** J Epidemiol 1994; 140: 585-97.


44.     Daniel L, Clarke-Pearson M, and Yousif D. **Green's Gynecology: Essentials of Clinical Practice**; 4th ed., Little, Brown Company; 1990; pp. 531-541.

45.     Wong C, Hempling RE, Piver MS, et al. **Perineal Talc exposure and subsequent epithelial ovarian cancer: A case control study.** Obstet Gynecol 1999; 93:372.

46.     Bristow RE and karlan BY. **Ovulation induction, infertility, and ovarian cancer risk.** Fertil steril 1996; 66:499.

47.  Rodrigue ZC, Tatham LM, Calle EE, et al. **Infertility and risk of fetal ovarien cancer in a prospective cohort of U.S. women Cancer control** .Cancer Causes Control ; 1998 ;9 :645.

48.  Rossing MA, Daling JR, Weiss NS, et al **.Ovarian tumor in a cohort of inferite women.** N Engl J Med.  1994; 331: 771.

49.  **"General Information about Ovarian Epithelial Cancer" NCl Cancer**. Gov. web site (Http://www.cancer.gov.).

50.  Smith LH and Oi RH. **Detection of malignant ovarian neoplasm's: A review of the literature I. Detection of the patient at risk, clinical, radiological and cytological detection.** Obstet Gynecol Surv.; 1984; 39:313.

51.  Pollock RE.   **Manual of Clinical Oncology**; 7$^{th}$ ed.; Wiley. Liss, Inc.; 1999; p.542.

52.  Campbell S, Bhan V, Royston P, et al. **Transabdominal ultrasound screening for early ovarian cancer.** BNJ 1989; 299:1363.

53.  Bast RC, Klug TL, St John E, Jenison E, Niloff JM, Lazarus H, Berkowitz RS, Leavitt T, Griffiths CT, parker L, zurawski VR, and knapp RC . **A radioimmunoassay using a monoclonal antibody to monitor the course of epithelial ovarian cancer.**   N  Eng J Med 1983; 309:883-887.

54.  Schwartz PE, and Taylor Kjw. **Ovarian Cancer: Epidemiological perspectives with Developments in Early Diagnosis**; the Parthenon publishing group New York; 1994, p. 257.

55.  James B, Wyngaarden Lloyd H, and smith JR, **Claude Bennett Cecil: textbook of Medicine**. 20$^{th}$ ed.; W.B. Saunders Company; 1996; pp. 1021-1022.

56.  Celluzzi CM, Mayordomo CI, Storkns WJ, Lotze MT, and Falo LD. **Peptide-pulsed dentritic cells induce antigen-specific CTL-mediated protective tumor immunity**. J Exp Med 1996; 183:283.

57.  Kawashima I, Hudson SJ, Tsai V, Southwood S, Takesako K, Appella E, Sette A, and Celis E. **The multiepitope approach for immunotherapy for**

cancer: identification of several ETL epitopes from various tumor associated antigens expressed on solid epithelial tumors. Hum Immunol 1998; 59:1.

58.　　　Vogle FD, Strickeler E, Weyermann M, Kohler T, Gill H, Negri G, kreienberg R, and Runnebaum IB . **p53 auto antibodies in patients with primary ovarian cancer are associated with higher age, advanced stage and higher proportion of a p53-posative tumor cell.** Oncology; 1999; 57:324.

59.　　　Vikhanskaya F, D'Incalci, and Broggini M. **p73 competes with p53 and attenuates its response in a human ovarian cancer cell line.** Nucleic Acid Res 2000; 28: 513.

60.　　　Werness BA, Freedman A, Piver MS, Romero-Gutierrez M, and petrow E. **Prognostic significance of p53 and p21 (waf1/cip1) immunoreactivity in epithelial cancer of the ovary.** Gynecol Oncol 1999; 75:413.

61.　　　Presneau N, Laplace-Marieze V, sylvain V, Lortholary A, Hardouin A, Bernard- Gallon D, and Bignon Y. **New mechanism of BRAC1 mutation by deletion / insertion at the same nucleotide position in three unrelated fresh breast cancer.** Hum. Genet 1998; 103:334.

62.　　　Lancaster JM, Garney M, and Futreal PA. **BRAC1 and 2: A genetic like to familial beast and ovarian cancer.** Medscape women's Health 1997; 2:7.

63.　　　Woolas RP, xu FG, Jacobs IJ, Yu YH, Daly L, Berchuck A, Soper JT, Clarke- pearson DL Oram DH, and Bast RC. **Elevation of multiple serum markers in patients with stage I ovarian cancer.** J Natl Cancer Inst 1993; 85:1748.

64.　　　kufe D, Inghirami G, Abe M, Hayes D, Justi-wheeler H, and schlom J. **Differential reactivity of a novel monoclonal antibody (DF3) with human malignant versus benign breast tumors .** Hybridoma 1984; 3:223.

65.　　　Mckenzie SJ, Desombre KA, Bast BS, Hollis DR, Whitaker RS, Berchuck A, Boyer C M, and Bast RC. **Serum levels of HER-2 neu(C-erbB-2) correlate with over expression of p158 neu in human ovarian cancer.** Cancer

1993; 71:3942.

66.	Hancock MC, langton BC, Chan T, Toy P  Monahan JJ, Mischak RP, and Shawver L K. **A monoclonal antibody against the c-erbB-2 portion enhances the cytotoxicity of cis-diaminedichloro-platinum against human breast and ovarian tumor cell lines.** Cancer Res 1991; 51:4575.

67.	Alexander WK and William RH. **Ovarian papillary serous tumors of low malignant potential (serous borderline tumors). A long term follow-up study, including patients with micro invasion, lymph node metastatasis, and transformation to invasive serous carcinoma.** Cancer 1996; 78:278.

68.	Thigpen JT, Lambuth BW, and Vance RB. **Management of stage I and II ovarian cancer.** Semin Oncol 1991; 18:596.

69.	Kristensen GB and Trope C. **Epithelial ovarian carcinoma.** Lancet 1997; 349:113-117.

70.	Kumar P, Rehani MM, Kumar L, Sharma R, Bhatla N, Chaudharg R, Thulkar S, Sunderam KR, and Kumar N. **Tumor marker CA125 as an evaluator and response indicator in ovarian cancer : its quantitative correlation with tumor volume.** Med Sci Monit 2005; 11: CR 84.

71.	Colakovic S, Lukic V, Mitrovic L, Jelic S, Susnjer S, ,and  Marinkovic J. **Prognostic value of CA125 kinetics and half time in advanced ovarian cancer .** Int J Biol. Marker 2000; 15:147-152.

72.	Hempling RE, Piver MS, Natarajan N, Baker TR, Thompson JM, Hicks ML, and Metlin CJ. **Predictive value of serum CA125 following optimal cytoreductive surgery during weekly cisplatin induction therapy for advanced ovarian cancer.** J of Surg Onco 1993; 54:38-44.

73.	Jacobs IJ and Bast RC. **The CA125 tumor –associated antigen: a review of the literature.** Human Reproduction; 1989; 4:pp.1-12.

74.	Backston T, Mahlck CA, and Kjellgrea O. **Progesterone as a possible tumor marker for nonendocrine ovarian malignant tumors.** Gynecol Oncol 1983; 16:129.

75. Heinnon PK, Tuimala R, Pyy KK, and Pystyam P. **Human placental alkaline phosphatase in benign and malignant ovarian neoplasia.** Br J Obstst & Gynecol 1982; 89:84.

76. Tholander B, Taube A, Lingew A, sjoberg O, Stendahi U, and Tamsen L. **Pretreatment serum level of CA125, CEA, tissue polypeptide antigen and placental alkaline phosphatase in patient with ovarian carcinoma.** Gynecol Oncol 190; 39:26-33.

77. Shabana A, Onsrud M. **Tissue polypeptide-specific antigen and CA125 as serum tumor markers in ovarian carcinoma.** Tumor Biol 1994; 15:361.

78. John R, Van Nagell JR, Pletsch A, and Goldenberg M. **A study of cyst fluid and plasma carcinoembrionic antigen in patients with cystic ovarian neoplasm's.** Cancer Res 1975; 35:1433-1437.

79. Negishi Y, Iwabuchi H, Sakunaga H, et al. **Serum and tissue measurement of CA72-4 in ovarian cancer patients.** Gyncol Oncol 1993; 48:149-54.

80. Berek JS and Martinez-Maza O. **Molecular and biological factors in the pathogenesis of ovarian cancer.** J Reprod Med 1994; 39:241-248.

81. Schwartz FE, Chambers SK, Chambers JT, Gutman J, Katopodis N, and Foemmel R. **Circulating tumor markers in the monitoring of gynecological malignancies** Cancer 1987; 60:353-61.

82. Berek JS, and Bust RC. **Ovarian cancer screening "the use of serial complementary tumor markers to improve sensitivity and specificity for early detection"** Cancer 1995; 76:2092-96.

83. Hanisch FG, and Dienst C. **CA125 and CA19-9: two cancer associated sialylsaccharide antigens on a mucus glycoprotein from human milk.** Eur J Biochem 1985; 149:323-330.

84. Sekine H, Ohno Tand Kufe DW. **Purification and characterization of a**

high molecular weight glycoprotein detected in human milk and breast carcinomas. J Immunol 1985; 135:3610-3615.

85. Shimizu M, and Yamauchi K. **Isolation and characterization of mucn like glycoprotein in human milk fat globule membrane**. J Biochem 1982; 91: 515-524.

86. Tsubura A, Morii S, Vdea S, Sasaki M, Zother S , Waltzing V, Mooi W, Hageman PC, Hilkens J, and Tweel JV. **Immunohistochemical demonstration of MAM-3 and MAM-6 antigen in normal human skin and their tumors.** Arch Dermatol Res. 1987; 279:550-557.

87. Sekin H, Hayes DF, Ohno T, et al. **Circulating DF3 and CA125 antigen levels from patients with epithelial ovarian carcinoma.** J Clin Oncol 1985; 3:1355-63.

88. Harbest AL. **The epidemiology of ovarian carcinoma and the current status of tumor markers to detect disease.** J Obstet Gynecol 1994; 170:1099-107.

89. Berek JS, Chung C, kaldi K, Watson JM, Knox RM, and Martinez Maza O. **Serum interleukin-6 levels correlate with disease status in patients with epithelial ovarian cancer.** J Obstet Gynecol 1991; 164:1038-43.

90. Gotlib WH, Abrams JS, Watson JM, Velu T, Martine Z, Mazo O, and Berek JS. **Presence of interleukin 10 (IL 10) in the ascites of patients with ovarian and other intra abdominal cancer.** Cytokine 1992; 4:385-90.

91. Ramakrishhan S, Xu FJ, Brandt SJ, Niedel JF, Bast RC, and Brown EL **.Constitutive production of macrophage colony-stimulating factor by human ovarian and breast cancer cell lines**. J Clin invest 1989;83:921-926.

92. Xu FJ, Ramakrishhan S, Daly L, Soper JT, Berchuck A, Clarke PD, et al. **Increased serum levels of macrophage colony stimulating factor in ovarian cancer.** Is J Obstet Gynecol 1991; 165:1356-62?

93. Xu FJ, Yu YH, Daly L, Desombre K, Anselmino L, Hass GM, et al. **The**

OVX1 radioimmunoassay complements CA125 for predicting the presence of residual carcinoma at second-look surgical surveillance procedures. J Clin Oncol 1993; 11:1506-11.

94.     Knauf S, Anderson DJ, Knapp RC, and Bast RC. **A study of NB/70K and CA125 monoclonal antibody radioimmunoassay for measuring serum.** Am J Obstet Gynecol 1985; 152(7):911-913.

95.     Schwartz PE, Chaambers JT, Taylor KJ, et al. **Early detection of ovarian cancer: preliminary results of the Yale Early Detection Program Yale.** J Biol Med 1991; 64:573-82.

96.     Soper JT, Hunter VJ. Daly L, Tanner M, Creasman W, and Bast RC. **Preoperative serum tumor-associated antigen levels in women with pelvic masses.** Obstet Gynecol 1990; 75:249.

97.     Davis HM, Zurawski VR, Bast RC, and Klug TE. **Characterization of the CA125 antigen associated with human epithelial ovarian carcinomas.** Cancer Res 1986; 46:6143-6148.

98.     Nagata A, Hirota N, Sakai T, Fujimoto M, and Komoda T. **Molecular nature and possible presence of a membranous glycophosphatidylinositol anchor of CA125 antigen.** Tumor Biol 1991; 12:279.

99.     Bast RC, Boyer JI, Xuf J, Wiener J, Kohler M, and Berckuck A. **Cell growth regulation in epithelial ovarian cancer.** Cancer 1993; 71:1597-1601.

100.    O'Brien TJ, Beard JB, Underwood LJ, and Shigemasa K. **The CA125 gene: A new discovered extension of the glycosylated N-terminal domine doubles to size of this extracellular superstructure.** Tumor Biol 2002; 23: 154-169.

101.    O'Brien TJ, Beard JB, Underwood LJ, Dennis RA, Santin AD, and York L. **The CA125 gene: an extacllular superstructure dominated by erepear sequences.** Tumor Biol 2001; 22:348-366.

102.    Kui Wong N, Easton RL, Panico M, Sutton Smith M, Morrison JC,

Lattanzio FA, Morris HR, Clark GF, Dell A, and Patankar MS. **Characterization of the oligosaccharides associated with the human ovarian tumor marker CA125.** J Biol Chem 2003; 278: 28619-28634.

103.     Yin BW and Lloyd KO. **Molecular cloning of the CA125 ovarian cancer antigen identification as a new mucin, MUC16.** J Biol Chem 2001; 276, 27371-27375.

104.     Wreschner DH, McGuckin MA, Williams SJ, Baruch A, Yoeli M, Ziv R, Okun L, ZareTsky J, Smorodinsky N, Keydar I, Neophytou P, Stacey M, Lin HH, and Gordon S. **Generation of ligand-receptor alliances by "SEA" model-mediated cleavage of membrane-associated mucin proteins.** Protein Sci 2002; 11:698-706.

105.     Hardardottir H, Parmely TH, Quirk JG, Sanders MM, Miller FC, and O'Brien TJ. **Distribution of CA125 in embryonic tissue and adult derivation of the fetal periderm.** J Obstet Gynecol 1990; 163:1925-1931.

106.     Fukuda M, **Roles of mucin-type O-glycans in cell adhesion.** Biochim Biophys Acta 2002; 1573:394-405.

107.     Bresalier RS, Byrd J, Wang L, and Raz A. **Colon cancer mucin: A new ligand for the beta-galactoside binding protein galactin-3.** Cancer Res 1996; 56: 4354-4357.

108.     Seelenmeyer C, Wegehingel S, Lechner J, and Nickel W. **The cancer antigen CA125 represents a novel counter receptor for galactine-1.** J Cell Science 2003; 116:1305-1318.

109.     Perillo NL, Marcus ME and Baum LG. **Galectins: versatile modulators of cell adhesion, cell proliferation, and cell death.** J Mol Med 1998; 76: 402-412.

110.     Gaetje R, Winnekendonk DW, Scharl A, and Kaufmann M. **Ovarian cancer antigen CA125 enhances the invasiveness of the endometriotic cell line EEC145.** J Soc Gynecol Investig 1999; 6: 278-281.

111. Rump A, Morikawa Y, Tanaka M, Minami S, Umesaki N, Takeuchi M, and Miyajima A. **Binding of ovarian cancer antigen CA125/MUC16 to mesothelin mediates cell adhesion.** J Biol Chem 2004; 279(10): 9190-9198.

112. Hassan R, Bera T, and Pastan I. **Mesothelin: A new target for immunotherapy.** Clin cancer Res 2004; 3937-3942.

113. Bast RC, Siegal FP, Runowiecz C, Klug TL, and Knapp RC. **Elevation of serum CA125 prior to diagnosing of an epithelial ovarian carcinoma.** Gynecol Oncol 1985; 22:115-120.

114. Bon GG, Kenemans P, Verstraeten R, Van kamp GJ and Hilgers J. **Serum tumor marker immunoassay in gynecologic oncology : establishment of reference values.** J Obstet Gynecol 1996 174:107-114.

115. Lloyd KO and Yin BW. **Synthesis and secretion of the ovarian cancer antigen CA125 by the human cancer cell line NIH: OVCAR-3.** Tumor Biol 2001; 22: 77-82

116. Meyer T, and Rustin GJ. **Role of tumor markers in monitoring epithelial ovarian cancer.** Br J Cancer 2000; 82:1535-1538.

117. Bast RC, Feeney M, Lazarus H, Nadler LM, Colvin RB, and Knapp RC. **Reactivity of monoclonal antibody with human ovarian carcinoma.** J Clin Invest 1981; 68: 1331-1337.

118. Nustad, et al. **Specificity and affinity of 26 monoclonal antibodies against the CA125 antigen: first report from the ISOBM TD-1 workshop.** Tumer Biol 1996; 17:196-219.

119. Lehtovirta P, Apter D, and Stenman VH. **Serum CA125 levels during the menstrual cycle.** Br J Obstet Gynecol. 1999; 97: 930-933.

120. Bonfrer JMG, Korse CM, Verstraeten RA, Van Kamp GJ, Hart AAM, and Kenemans. **Clinical evaluation of the BYK LIA-mat CA125 II assay: Discussion of a reference value.** Clin Chem 1997; 43: 491-497.

121.    Guadagni F, Marth CH, Zeimet AG, Ferroni F, Spila A, Abbolito R, Roselli M, Greiner JW, and Schlom J. **Evaluation of markers in patients with gynecologic diseases.** AMJ Obstet Gynecol 1994; 171:1183-91.

122.    Austoker J. **Screening for ovarian, prostatic and testicular cancer.** Br Med J   1994; 309:315-320.

123.    Monagham JM. Malignant diseases of ovary. Dewhursts text book of Obstetrics and Gynecology for postgraduates, 6 editions; Blackwell science Ltd. 1999; PP.590-592.

124.    Zurawski VR, Orjaseter H, Andersen A, et al. **Elevated serum CA125 levels prior to diagnosis of ovarian neoplasia: relevance for early detection ovarian cancer  Int J** Cancer 1988; 42:677.

125.    Roupaz FE, Raftopoulos V, Tzavelas G, Kotrotsiou E, Sotiropoulou P, Karanikola E, Skifla, and Ardavanis A. **Serum CA125 combined with transvaginal (TSV) ultrasonography for ovarian cancer screening.** Invivo 2004;18(6):831-836

126.    Bast RC, Xu F, Woolas RF, Yu Y, Conaway M., O' Briant K, et al. **complementary and coordinate markers for detection of epithelial ovarian cancers,** In: sharp F., Mason P, Blackett T, and Berek JS, editors. Covarian Cancers 3; Chapman and Hill; London; 1995; P.P. 189-192.

127.    Schutter EMJ, Kenemans P, Sohn C, Kristen P, Crombach G, Westermann R et al. **Diagnostic value of pelvic examination, ultrasound and serum CA125 in post-menopausal women with pelvic mass.** Cancer 1994; 74:1398-1406.

128.    Rustin GJS. **The clinical value of tumer markers in the malignant of ovarian cancer.** Ann Clin Biochem 1996;33:284-289.

129.    Wagner U, Kohler S, Prietl G, Giffels P, Schmidt-Nicolai S, Schlebusch H, et al. **Monoclonal antiidiotypic antibodies in immunotherapy of ovarian carcinoma (MAB CA125) and breast carcinoma** .Zentrabl Gynakol 1999; 121:190.

130. Wagner U, Schlebusch H, Kohler S, Schmolling J, Grunn U, and Krebs D. **Immunological responses to the tumor associated antigen CA125 in patient with advanced ovarian cancer induced by the murin monoclonal anti-idiotypic vaccine A CA125.** Hybridoma 1997; 16:33-40.

131. Nusted K, Lloyd KO, Nilsson O, and O'Brient TJ. **Epitopes on CA125 from cervical mucus and ascites fluid and characterization** Tumer Biol 2002; 23:303-314.

132. Einhorn N, Sjovall K, Knapp RC, Hall P, Scully RE, Bast RC, and Zurawski VR. **Prospective evaluation of serum CA125 levels for the early detection of ovarian cancer.** Obstet Gynecol 1992; 80: 14-18.

133. Boerman OC, Thomas GMG, Segers MFG, Kenemans P, Lovgren, Zurawski VR, Haisma HJ, and Poels LG. **Time-resolved immunoflurometric assay for the ovarian carcinoma-associated antigenic determinant CA125 in serum.** Clin Chem 1987; 33(12): 2191-

134. Schollerr N, Crawford M, Sato A, Drasche CW, O'Biant KC, Kiviat N, Andrson GL, and Urban N. **Bead-based ELISA for valedation of ovarian cancer : Erley detection    markers** . Clin Cancer  Res 2006 ;12:2117-2124.ancer

135. **Bast RC and Knapp CC.  CA125: History, current status, and future prospects. MJM**, 1997; 3:67-71.

136. Lowry OH, Rosenrough NJ, Farr AL, and Randel RJ. **Protein measurement with folin phenol reagent.** J Biol Chem 1951; 193: 365-375.

137. Fleuren GJ, NAP M, Aalders JG, Trimbos JB, and DE Bruijn NWA. **Explanation of the limited correlation between tumor CA125 content and serum CA125 antigen levels in patients with ovarian tumors.** Cancer 1987; 60: 2437-2442.

138. Al-Barazanji AK. (2002) **"The accuracy of malignant risk index based onCa125, ultrasound and menopausal state".** Thesis, supervised by Kais Kubba and Suhail Najim Al-Salam, submitted for the degree of fellowship of

Arab Board of Obstetrics and Gynecology.

139.        Malkasion G.D. Jr., Knapp R.C., Lavin ph. T., Zurawski V.R. Jr., Podratz K.C., Stonhope CR, Mortel R, Berck JS, Bast RC, and Ritts RE. **Preoperative evaluation of serum CA125 levels in pre-menopausal and post-menopausal patients with pelvic masses: Discrimination of benign form malignant disease. J Obestet Gynecol** 1988; 159:341-6.

140.        James T Wu, Terry M, Joseph AK, and David PK. **Improved specificity of the CA125 Enzymeimmunoassay for ovarian carcinomas by use of the ratio of CA125 to carcinoembryonic antigen.** Clin Chem 1988; 34/9: 1853-1857.

141.        Nagell JR, Meeker WR, Parker J., and Harraison JD. **Carcinoembryonic antigen in patients with gynecologic malignancy.** Cancer 1975; 35: 1372-1376.

142.        Nagell JR, Donaldson ES, Gay EC, Sharkey RM, Rayburn P, and Goldenberg DM. **carcinoembryonic antigen in ovarian epithelial cystadenocarcinomas. Cancer** 1978; 41: 2335-2340.

143.        Donaldson ES, Nagell JR, Pursell S, Gay EC, Meeker WR, Kashmiri R, and Voorde J. **Multiple biochemical marker in patients with gynecologic malignancies.** Cancer 1980; 45: 948-953.

144.        Freifrlder D. **"Physical Biochemistry: Application to Biochemistry and Molecular Biology"**; 2nd ed.; San Francisco: W.H. Freeman & Company. 1982; Chapter 14; pp. 494-591.

145.        Changux JP. **Responses of actylcholinesterase from torepedo marmorata to salts and curarizing drugs.** Mol. Pharmacol 1966, 2:369.

146.        Helen CH, Mansel H, Siraj M, and Niel S. **Essential of Clinical Immunology**; 4th ed.; London Blackwell Science Ltd; 1999; Chapter 19:pp.314-321.

147.        Clackson T, Hoogenboon HR, Griffiths AD, and Winter G. **Marking

**antibody fragments using phage display libraries.** Nature 1991; 352:624-628.

148.     Dixon M, and Webb E, **Enzymes**; 3$^{rd}$ ed.; London; Longman Group Limited; 1979; pp.273.

149.     Devlin TM. **Text Book of Biochemistry with Clinical Correlation**; 2$^{nd}$ Ed; John Wiley and Sons Inc.; New York; 1986; pp.125-66.

150.     Melander W and Horvath C. **Salt effect on hydrophobic interactions in precipitation and chromatography of proteins: an interpretation of the lyotropic series.** Arch Biochem Biophys 1977; 183: 200-215.

151.     Collins KD. **Charge density-dependent strength of hydration and biological structure**. Biophys J.; 1997; 72: 65-76.

152.     Evans JS and Levine BA. **Protein–protein interaction sites in the calcium modulated skeletal muscle troponin complex.** J Inorg Biochem 1980; 12:695.

153.     Jones S, and Thornton JM. **Principles of protein –protein interactions.** Proc Natl Acad Sci 1996; 93:13-20.

154.     O'Brien TJ,   Hardin JW, Bannon  GA, Norris JS, and Quirk JG. **CA125 antigen in human amniotic fluid and fetal membranes.** Am J Obstet Gynecol 1986; 155(1): 50-55.


155.     Scopes RK, **Protein Purification Principles and Practice**; New York; Springer Verlag. 1982; pp. 162-197.

156.     Price NC, and Stevens L. **Fundamentals of Enzymology**; 2$^{nd}$ ed.; Newyork, Oxford University Press; 1986; pp. 125.

157.     Ormerod MG, Steel K, Westwood JH, and Mazzini MN. **Epithelial membrane antigen: partial purification, assay and properties.** Br J Cancer 1983; 48:533-541.

158.     Segal I.H.; **Biochemical Calculations**; 2$^{nd}$ ed.; John Wiley and Sons; 1976; pp. 278-373.

159.     Haisma HJ, Battaile A, Stradtman EW, Knapp RC, and Zurawski VR.

Antibody antigen complex formation following injection of CA125 monoclonal antibody in patients with ovarian cancer. Int J Cancer 1987; 40: 758-762.

160.     Wiseman T, Williston S, Randts J, and Lnng-Nam Lin. **Rapid measurement of binding constants and heats of binding using a new titration calorimeter.** Anal Biochem 1989; 179:131-137.

161.     Rosier JS, Gokulrangan G, Girault H, Sovojanovsky S, and Wilson GS. **Characterization of protein adsorption and immunosorption kinetic in photoabelated polymer microchanals.** Langmuir 2000; 16: 8489-8494.

162.     Weiland GA, Minneman KP, and Molinoff PB. **Thermodynamic of agonist and antagonist interactions with mammalian β adrenergic receptors.** Mol. Pharmacol 1980; 18: 341.

163.     Camacho Cj, Weng Z, Vajda S, and Delisi C. Free energy landscapes of encounter complexes in protein-protein association. Biophys J 1999; 76:1166-1178.

164.     Seeley DH, Wang WY, and Salhanick HA. Temperature dependence of kinetic interactions between progesterone and uterine cytoplasmic receptor. **Biochem Biophy Acta** 1980; 632: 536-543.

165.     Forde A, and Coley J. Choosing and characterizing antibodies. In: Goaling JP. Editor. Immunoassays A practical approach ; Oxford university press London; p.p.62-63

166.     Weiland GA and Molinoff PB. **Qutitative analysis of drug-receptor interaction I. Determination of kinetic and equilibrium properties.** Life Science; 1981; 29: 314.

167.     Nemethy G and Scherag AJ. **The structure of water and hydrophobic bonding in proteins: III the thermodynamic properties of hydrophobic bond in protein.** Phys Chem 1962; 66:1775.

168.     Waelbroeck M, Van Obberghen E, and De Meyts p. **Thermodynamics of the interaction of insulin with its receptor.** J Biol Chem 1979; 259:7736.

169. Haro LS, and Talamantes FJ. **Thermodynamics and kinetics of mouse prolactin-hepatic receptor interaction.** Mol Cell Endocrinol 1985; 43:199

170. Ross PD and Subramanian S. **Thermodynamics of protein association reaction: Forces contributing to stability.** Biochemistry 1981; 20:3096.

171. Blumenthal DK and Stull JT. **Effect of pH, ionic strength, and temperature on activation by calmodulin and catalytic activity of myosin light chain kinase.** Biochemistry 1982; 21:2386-2391.

172. Laport DC, Wireman EM, and Storm DI. **Calcium-induced exposure of a hydrophobic surface on calmodulin.** Biochemistry 1980; 19: 3814.

173. Johnstone A and Thorpe R. **Immunochemistry in Practice**; 3$^{rd}$ ed.; Blackwell Science Ltd.; 1996; p.p. 1-4, 292-311.

174. Bujalowski W and Jezewska MJ. Quantitative determination of equilibrium binding isotherms for multiple ligand-macromolecule interactions using spectroscopic methods. In: Michael G. Spectrophotometry and spectrofluorimetry: a practical approach .New York: Oxford; 2000; pp. 141.

175. Nolta K and Steck. **Isolation and initial characterization of the bipartite contractile vacuole complex from dictyostelium discoideum.** J Biol Chem 1994; 269:2225.

176. Scheraga HA. **Protein Structure**; New York: Academic Press; 1961; pp. 365-571.

177. Kiernan JA. **Histological and Histochemical Methods Theory and Practice**; 3$^{rd}$ ed., Reed Educational and Professional Publishing Ltd.; 1999; Chapter 19: pp. 391-398.

178. Williams CA and Chanse MW. **Methods in immunology and immunochemistry**; New York., Academic Press; 1968; vol II; Chapter10: pp 163-174.

179. Haider TM. (2004) **"Development of Radio Receptor Technique for Measurement of CA125 in Malignant and Benign Breast Tumors"**. Thesis, supervised by Al-Mudhaffar S.A., College of Science, Baghdad University.

180.     Al-Jobory E. (2004) **"Biochemical Characterization of CA125 in Sera and Tissue of Some Colorectal Tumors"**. Thesis, supervised by Al-Mudhaffar S.A., College of Science, Baghdad University.

181.     Mathews Ch k, and Holde KE. **"Biochemistry''** California the Benjamin /Cummings Publishing Co.; 1990; Chapter 6: pp. 191.

182.     Sheinerman FB, Norel L, and Honig B. **Electrostatic aspects of protein-protein interaction.** Curr Opin Struct Biol 2000; 10:153-159.

183.     Nils H Axelsen. **"Hand book of immuno precipitation"-in Gel Techniques**; 3$^{rd}$ ed.; WA. Benjamin Inc. London; 1983.

184.     Nagacura S, and Baba H. **Dipole moment and near ultraviolet absorption of some monosubstituted benzenes: The effect of solvent and hydrogen bonding.** Am Chem. Soc 1952; 74:5693.

185.     Pimentel GC. **Hydrogen bonding and electronic transitions: The role of the Franck-Condon principle. J Am Chem Soc** 1957; 79:3323.

186.     Silvestien. RM, Bassalar GC, and Marril. TC. **"Spectrophotometric identification of organic compounds"**; New York: John Wiley and Sons; 1981; pp.181.

Leach SJ. **physical principles and techniques of protein chemistry**; New York: Academic Press;

www.ingramcontent.com/pod-product-compliance
Lightning Source LLC
Chambersburg PA
CBHW080010210526
45170CB00015B/1969